共·享

——设计师的人文思考

SHARE - THE DESIGNER'S HUMANISTIC THINKING

邵唯晏　卜天静　编著

江苏凤凰科学技术出版社

前言

共享——未来

近几年来，大家在不断谈论“共享”理念，出现了许多以“共享经济”为口号的机构。共享经济本来是指把闲置的物件或劳动时间拿出来共享，平台本身只负责规则的制定、管理和执行，从中撮合合作（交易）的达成，但自身并不参与实际交易。但现在许多机构却打着“共享”的旗号，亲自主导并介入交易与运作，这样就违背了“共享”的初衷。共享的本质，在于凭借互联网的基础，以去中心化的链接，构建一种更开放、更具包容性的合作架构，以便寻找各种可能性来面对严峻的未来环境。

我们的下一代正面对着极为严峻的考验：气候的异常、资源的浩劫、极速的生活节奏、少子化的竞争、破碎化的社交……共享的机制或许有机会在这极端的环境中找寻可能的出路。日本著名的管理学家、经济评论家大前研一（Kenichi Ohmae）预测：未来因为 M 型社会的冲击，设计师将被迫转变为社会型设计师 (Social Inventor)，而共享将是一条重要的出路。因此，本书集结了新锐设计师、设计专业学生和学术界的老师，从多元角度来探讨共享未来的可能。本书同时也记录了一些实际的装置（Installation），它们已在世界各地小规模运作了，期望通过这种由下而上的力量来引导更多人的关注。

这几年，我在几个身份中不断切换，台湾中原大学建筑系及室内设计系的毕业班导师、竹工凡木研究室的设计总监和台湾“交通大学”的建筑博士生，不同身份的视角及相互依存共享的关联，让我有更多的机会来看待设计未来的多元发展。我发现“共享思维”的口号，让设计圈产生了根本上的变化。当信息扁平与知识共享充斥着我们的生活时，在看似极度的便利中，我们是否感受到正在逐渐失去什么？是对于信息的珍惜，还是对于物件的惜福，抑或是对于人与人之间的情感？所以共享不能是口号，应真正串联起来变成一股正向的系统来衔接未来！

本书的出版也谢谢中国台湾逢甲大学的陈文亮主任及大陆的卜天静、田甲、孙浩晨三位亦生亦友的知己，还有我在马来西亚和日本的学生们，感谢译者王先志，在大家点滴的积累下，或许对于未来只是些许的改变，但我相信“共享”的机缘与力量会继续牵连着彼此，持续蔓延，直到未来。

Preface

Share and Future

Recently, everyone constantly discusses about " Share " concept. It appears many organizations of Sharing economy. Actually, everyone doesn't know sharing economy is sharing un-used object or working hours. This platform is responsible for the rules, management and execution which deals with cooperation and doesn't involve in real transaction. Nowadays, many organizations have the main idea of " Share" and intervene the bargain and operation by themselves. It leads to lack of the original sharing. The essence of share is rely on the basic interconnection network, the link of decentralization, and construct one kind of inclusive cooperation structure which is finding any kinds of possibilities to face the severe future environment.

Our generation confront the acid test, climate change, resources depletion, quick pace of life, low birth rate, fragmented social interaetion. Maybe the strategy with extreme environment sharing will find the way to go. Kenichi Ohmae, Japanese management expert and economist critic predicts the effect M-shaped Society in the future. Designers will change to Social Inventor. And share will be an important way to go. Thus, this book collects new generation designers, students, and academic professors to discuss the possibility of the future in diversity. At the same time, it records some practical installation and operation from all over the world and expect the force from down to up, then let many people to pay attention on it.

I have multiple roles in the graduating class professor of Architecture and Interior Design in Chung Yuan Christian University, the CEO of CHU-studio, the Ph.D student of Chiao Tung University. Different roles and viewpoints are connected to share to each other. It gives me many opportunities to view the develop of design. I find that we follow share concept which makes metamorphosis in design circle. To information flatting and Knowledge sharing including in our life, it seems like convenient , but lack of something. To cherish the information and things, to get close to each other? Share isn't a slogan and that is the way to connect with the future.

I want to say thank you for the director of Architecture Wen-Liang Chen in Feng Chia University. Furthermore, Sissi Bu, Tian Jia, and Sun Howchen as my best friends. Otherwise, my students are from Mainland China, Hong Kong, Macau, Malaysia and Japan. Thanks to the translator Mr Wang Xianzhi. Due to everyone's effort, it may change some situation to the future, but I believe the opportunities and strength of " Share" will extend until the future.

目录

第 2 章　SLRD 2017 参展作品

Chapter 2　SLRD 2017 EXHIBITION WORKS　061

第1章
CHAPTER 1

大家论共享
DISCUSS SHARING

陈文亮　台湾
逢甲大学建筑专业学院助理教授
室内设计学士学位学程主任

共享——与设计教育及设计人的对话

在公园儿童游戏场边总能听到家长对小孩说，要与弟弟妹妹或哥哥姐姐们“共享”，可见基本的“共享”概念其实很早就在我们的生活教育中被清楚地实践着，但是为什么这么好的观念竟在我们成长受教后渐渐地流失了？这的确值得我们大家好好反省！

“共享”或“分享（Sharing）”是指资源或空间的分享及利用。如今全球社会变化的脚步加速，封闭式的学习方式在现今社会有着极大的争议性，设计教育在一个变化快速的网络信息时代重新自我检视，是绝对必要的。1971 年维克多 · 帕帕奈克出版了《为真实世界的设计》一书，强调设计师的社会责任；20 世纪 90 年代以后，住宅设计兴起发展；2000 年开始，设计观点侧重于它所创造的经济价值；到了 21 世纪，“为社会而设计”的意识再度被各界唤起。

过去的设计价值观念中，往往都只是为了追求产品对象的外在美观及其设计创意，借以引发消费者购买交易的欲望，而过去社会设计的主要目的在于满足社会基本的需求，强调为大多数公众而设计。然而，现今的社会设计是运用设计的思维去做一个在生活里真正需要的东西，设计能量展现与设计人的使命都攀向另一个高峰，而其内涵则提升到有其特定社会意义价值的具体行动、改革与服务。

20 世纪 60 年代开始，台湾建筑高等教育就是所谓的台湾地区“老六校”，分别是成功大学（台南）、东海大学（台中）、中原大学（中坜）、淡江大学（台北）、逢甲大学（台中）及中国文化大学（台北）六所。其中成功大学及中国文化大学是四年学制，其余四所大学都是五年学制。中国台湾建筑教育基本架构以欧美为学习对象，在信息闭塞与资源有限的年代，是以课堂及图书馆书籍为获取知识的重要途径，六校联谊虽然盛大但人数有限，教师轮岗带动不同学校之间的互动交流，学长学姐与学弟学妹之间的互动有生活及专业的传承……当时“共享”面向有限，却令人格外深刻和难忘。

现今台湾建筑设计、室内设计或空间设计专业有超过四十所大学或技术学院，毕业学分门槛相较于过去降低不少。必修学分降低，原本是让学生有更多的空间去开拓学习，然而被曲解的初衷成为学习怠惰的借口，教育松绑已强烈冲击了台湾高等教育的根基。也希望其他地方的高等教育能引以为戒。

最近十几年，两岸高等教育机构交流相对频繁，互动的时间及机会都增加了许多，其中设计教育间的交流也很多。我个人对两岸建筑或室内设计高等（本科生）教育上的心得就是：课程架构及教学模式上还可以再加强。现在大陆许多大学院校之间教学资源及学术之间的交流都让台湾高等教育界羡慕不已，可以通过更多实际有效的知识交流及经验共享来提升高等教育的竞争力。我坚信，设计教育不应只是单纯教育学生成为一个只会做设计的狭隘设计人，而是应该通过更多“共享”让自己的心胸、视野更加宽广；通过差异“共享”让设计面向更加丰富且多样；通过深刻“共享”让设计内涵更加具备温度及质感。“共享”能帮助设计教育培养出兼具质感与温度的设计人，相信“共享”也可以让两岸的设计教育理念及运作机制得到取长补短的机会，引领设计产业迈向另一个巅峰。

设计师真的是一个很特殊的身份，在某种程度上有时可以说是公众人物，同时也担负着一些社会责任。设计师不仅要完成业主想要的理想设计，也要帮助业主获得一分真实、舒适、健康、美观的空间享受，表现设计专业的同时还要能传递一种正确的生活方式和社会理念，能从有形的空间影响到无形的责任。

要通过设计工作的表现让更多社会上的“成功人士”来从事慈善公益事业，带动更多的人去关注社会公益，进而形成无限改变且利人利己的强大力量。学校在培养学生成为设计专业人才之外，也要引导教育学生培养“共享”关怀的胸怀，学校自身就可以身作则将教育事业与社会公益关怀相连接。

现在我以逢甲大学建筑专业学院为例，与大家分享培养设计人的社会责任及使命的经验。逢甲建筑专业学院现有建筑学士学位（五年制）、室内设计学士学位（四年制）及创新设计学士学位（四年制）三个专业设置，学生不仅能具备自己的专业技能，更能轻易地在这样的学制内拥有跨领域学习的机会，这便是一个广义“共享”概念下的教育方式。大一及大二采用不分流的教学方式让学生们互动学习，学习内容多样丰富，两年的建筑训练可以建构学生对建筑内外空间的理解。大三分流，着重为不同属性、不同专长的人提供深入学习发展的机会，协助学生打造可供自己憧憬挥洒的天空。

学校不仅为“共享”提供最好的教育资源与环境，更是有效结合学生与社会企业一起来承担更多的社会责任，比如我们要在偏僻的乡村建造一百座小书屋。逢甲大学建筑专业学院以一个设计人、一个设计专业教育单位的身份播下这颗社会公益的种子，让教学不只局限于教室里，让教学成为师生可以实际动手参与的过程，让教学成为共享资源、共享爱的过程。做公益的心当然也不局限在台湾，原本我们只是单纯地希望小书屋能够让需要的人使用，而现在的小书屋不仅激发了我们更多的社会责任感与公益活动的热心，也扩大了我们教学工作的视野，带来了更多的机会。没有共享助人的初衷是无法体会到实践之后的充实幸福的。

Communications among sharing, design educations and designers

It happens all the time that the children, in the park playground, are told to share with their siblings, which explains that the concept of sharing has been widespread since we were young. Nevertheless, the good tradition of sharing, as we grow up, is becoming obscure. We should work it out before it is getting worse.

The definition of "sharing," from Wikipedia, is the joint use of a resource or space. In the fast-growing world, the closed learning approach is controversial, and we should reexamine the design educations, making these compatible corresponding to the era of information and internet explosion. In 1971, Victor Papanek published the book *Design for Real World*, which advocates the importance of social responsibility for designers. The energy of design and mission of designers have reached their peaks, which means that we should create and provide the solid actions, resolutions and services symbolizing the genuine social values.

Since 1960s, the so-called old six higher-education of architecture schools in Taiwan are Cheng Kung University (in Tainan), Tunghai University (in Taichung), Chung Yuan University (in Chung Li), Tamkang University (in Taipei), Feng Chia University (in Taichung) and "Chinese Culture University" (in Taipei), among which the Cheng Kung University and "Chinese Culture University" have 4-year academic programs, and the others have the 5-year. The foundation of architectural educations in Taiwan mostly follow that in the Europe and U.S. In the era of the limited educational resources, the classes and libraries had become the crucial channel learning knowledge. Though the academic associations, among these 6 schools, were constrained, was not large in scale, the teacher exchanges and student communications were active and remembered.

At this moment, there are over 40 colleges having architectural, interior or space design programs. The required credits for graduation are becoming fewer, which implicitly make students disinterested in studying. The loosen education regulations and constant education reforms were intended to ereat more space for students to explore and study, but the mislinderstood intention impact on the environment of higher education in Taiwan, China. The lack of foresighted education policy in Taiwan, China is worrisome, which is taken as a warning for education in Mainland China.

The cross-strait academic exchanges have been more frequently in both time and opportunity, espeeially the education of design. My personal opinion on the cross-strait education of architecture and interior design is that the curriculum structures and teaching mode can be better. The development of design education has been progressing in recent years, nevertheless. Nowadays, the education resources and academic development in Mainland China are comparatively competitive with Taiwan. Making the national education competitive by means of intensifying the effective knowledge exchange and experience sharing is not easy, however, I firmly believe that design education should not be confined to teach student design, but develop the broader mind and the diversity by sharing. Sharing design makes design warm and elegant, and sharing design makes designers warm and elegant as well. I believe that sharing is an opportunity to make the cross-strait educations learn each other, especially sharing education concepts and operations, which make a better world for the design industry.

Designers are unique, some of whom are public figures with social responsibility. Designers not only present to clients the ideal designs, but also present to clients the real, comfortable, healthy and alluring space experiences. Acting professionally while expressing the correct life style and social missions are also part of the responsibility which is making the tangible space affect intangibly.

The design works would attract more "business elites" to engage in charity activities, which brings more attention to public welfare and turns the egocentric people into the altruistic. In additional to teaching students the knowledge of design, the schools have to guide the students to share and care. The schools, therefore, are being connected the social welfare as well.

Taking the School of Architecture in Feng Chia University as an example, I am sharing with you the experience for nurturing the social responsibility and mission for designers. In this school, there are three professional programs, incorporating the 5-year Bachelor of Architecture, 4-year Bachelor of Interior Design and 4-year Bachelor of Innovation Design, that provide specific and disciplinary educations. This is the example presenting the idea of sharing from the macroscopic perspective. For the first 2 years without academic majors, the freshman and sophomore students have diverse options in courses, which help the students to learn the fundamental knowledge of design. With requirements of majors in their third year, the students having different professions can have the in-depth training, making themselves prepare better for the future.

The School of Architecture in Feng Chia University not only provides students the better learning resources and environments, but also effectively integrates the power of alumni and corporations so as to take more social responsibilities. We have committed to creating 100 book houses for schools in the underprivileged minorities. Our school wants to act first as a designer that makes the education no longer limited to classrooms. The passion of public welfare is not confined to Taiwan. Primitively, we wanted to make the house books conveniently accessible for more people. At present, the house books not only awaken the consciousness of social responsibility and public welfare activity, but also broaden our horizon, making us realize that there is no fulfilling happiness until we learn to share and help.

邵唯晏　台湾
竹工凡木设计研究室主持人

共享——超扁平时代

时代的进程总是和科技的进步有着密切的关联。自从 1989 年互联网介入我们的生活，直到 2018 年当下互联网时代的全面发展，短短 30 年的时间造成了信息的极度扁平化和透明化，也因此造成了碎片时间（Fragmental Time）和流动空间（Liquid Space）的大量出现，使“共享”观念盛行。这也回应了村上隆（Murakami Takashi）所谈及的超扁平（Superflat）文化的崛起和托马斯 · 弗里德曼（Thomas Friedman）在《世界是平的》（*The World is Flat*）一书中所谈及的全球扁平化现象，并且建议活在当下的年轻人们，产品和生产在当代是平坦的，所以不要从事单一无创新的生产工作，要寻找各种可能的创意，才不会被发展中国家的廉价劳工或机器所取代。这种由下而上、强调共享互利、去中心化发展的网络（Net）表现，也全方面影响并介入了我们的生活。每个时代都有它的使命和标记，而当下的时

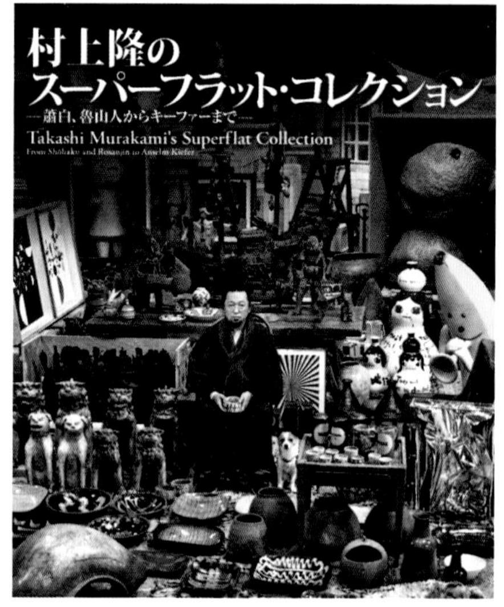

▲ 村上隆长期重视次文化，此为超平面收藏品特展。
Murakami Takashi always attaches importance to sub culture. This is the special exhibition of Superflat collection.

▲ 托马斯 · 弗里德曼的《世界是平的》一书，已出版到 3.0 版本，谈全球共享的扁平现象。
The World is Flat by Thomas Friedman has been published in 3.0 edition, discussing the flat phenomenon shared by the whole world.

空是“共享思维”的绝对主场。当前的共享经济有几个重要的特质：第一是建立在互联网基础上的信息平台；其二以闲置资源及资产（Idle Assets）使用权的暂时性转移为本质；第三是以物品的重复交易和高效利用为表现形式。2011 年美国《时代周刊》杂志将共享经济思维列入“改变世界的十大理念”之一。

先来点硬知识的定义，哈佛大学经济系教授 Martin Lawrence Weitzman 认为共享经济存在的形式基本上可分为三类：①基于共享和租赁的产品服务。这实际上是在同一所有者掌控下的特定物品在不同需求者间实现“使用权”的转移，如拼车网、房屋交换网等，而金融企业也是典型的例子，是一种基于分享经济理念的经济形态。②基于二手转让的思维让产品再流通。实质上是同一物品在不同需求者间依次实现“所有权”的转移。如 1999 年成立的克雷格列表网，这是一个通过大数据调研，不断反复运算更新的网上大型免费分类广告网站，如房屋租赁买卖、娱乐甚至是敏感的找寻异性朋友等服务。③基于资产和技能共享的协同生活方式或工作模式。无形资产指的是时间、知识和技能等，有形资产指的是空间、设备等，这种有形与无形资产交织共构所建立的“机制”，可供流动使用者快速便捷地加入与退出。比如共享办公室、众创空间、创客自造实验室等，这种微型尺度的作坊复制，类似 Airbnb 的模式，为在外出差者提供最佳的共享工作空间。此外，这种形式还包括一方利用空闲时间（无形资产）为另一方提供服务等形式，比如 Uber、滴滴等共享汽车平台等。

“共享”代表了思维解放，而互联网则成达到“共享”的手段。随着社交网络的日益成熟，共享在日常生活中已无处不在，当前共享内容不局限于虚拟资源，更是扩展到食、衣、住、行、教、乐等实体消费领域，进而形成了全新商业模式的共享经济。因此笔者试图通过经济的共享、建筑的共享、新零售的共享、沟通的共享四个方面来了解当下及探索未来。

▲ 最早由美国麻省理工学院发起的“数制工坊实验室”思潮，掀起全民共享 DIY 及自我创造的风潮。
FAB LAB trend of firstly initiated by MIT started a wave of universal DIY and self creation.

▲ Airbnb 的执行官布莱恩·切斯基（Brian Chesky）于旧金山的活动上阐述未来的共享机制。
Brian Chesky, Executive Director of Airbnb, is describing the future sharing mechanism in San Francisco.

Sharing – the era of Superflat

The time evolution has been always firmly connected to the technology growth. Since 1989 when the internet was launched till 2018 when it becomes universal, the information development has become utterly flat and transparent, the phenomenon contributing to the massive reiterations of "fragmental time" and "liquid space" while resulting in the popularity of the "sharing" concept. Corresponding to Murakami Takashi who coined the Superflat culture and to Thomas Friedman who described the incident of flattening world in his book *The World is Flat*, the flourishing concept of sharing reminds the youngsters that they live in the world with flattening products and manufactures, and that they should be more creative and competent compared to the cheap labors and machines in the developing countries. Furthermore, the advancement of the bottom-up, reciprocally beneficial and decentralized network has comprehensively affected our life. Every generation, therefore, has its specific mission and feature that in the present is the "sharing concept." Nowadays, the sharing economy has some imperative characteristics, of which the first is using the internet as the primary information platform. The second is the transitional relocation of idle assets; the last is the goods circular transactions and efficient utilizations. The America magazine "Time" in 2011, consequently, listed the sharing economy as one of the "10 ideas for the next 10 years."

Martin L. Weitzman, the professor of economics at Harvard University, categorized three essences of sharing economy, of which the first is the services of goods sharing and leasing. This defined the explicit goods of ownership shifting the right of use among users, e.g. carpooling and home exchange. The financial industry is a form of sharing economy as well. The second is the recirculation of products that transfer the ownership on the basis distinct needs. Craigslist, found in 1999, is a big classified website for advertisements offering the services according to the big data analysis. Its business incorporates the household sales and rentals, the entertainments and the friends. The last is either the sharing lifestyles or works depending on assets of skills. The assets can be either intangible encompassing the time, the knowledge and the skill, or tangible including the space and the equipment. Such "mechanism" constituted on the basis of the tangible and intangible assets is convenient for accessing and quitting. The instances encompass Coworking space, Makerspace, Fablab, Hackerspace and Techshop. Airbnb, a micro-scale workshop, is ideal for business travelers. Uber and carpooling represent the service of intangible assets, like time.

The idea of sharing signifies the rise of mind emancipation, and the internet is mandatory. Following the advanced development of social media, the activities of sharing is omnipresent in our life. The context of sharing, moreover, is extensively applied to the food, clothing, housing, transportation, education and entertainment. A new business model of sharing economy has been established, and I am uncovering the future evolution from the perspectives of sharing economy, sharing housing, sharing retail and sharing communication.

经济的共享

关键词：共享经济、众筹、社会机制

共享经济又称租赁经济，是一种共享人力与资源的新的社会运作方式。包括不同个人与组织对商品和服务的创造、生产、分配、交易和消费的共享。共享经济同时也是指拥有闲置资源的机构或个人有偿让渡资源使用权给他人，让渡者获取回报，分享者利用分享自己的闲置资源创造价值。共享经济的概念最早起源于马库斯 · 菲尔逊（Marcus Felson）与乔 · 思帕斯（Joe L. Spaeth）的论文《社区结构和协作消费：一种日常活动的方法》（*Community Structureand Collaborative Consumption: A Routine Activity Approach*）。该论文对汽车共享的行为进行了探讨，发现闲置资源其实可以通过使用权的“共享”来实现更有效的配置，这也是当今共享产业（如 Uber、Zipcar、Airbnb）最早的发展模型。

再者，根据共享经济的特性和目的，英国学者瑞秋 · 波茨曼（Rachel Botsman）和美国传奇投资人吉姆 · 罗杰斯（Jim Rogers）将此概括为四大类别：共享消费、共享生产、共享学习及共享金融。前三者比较好理解，但谈到共享金融就更有趣了，因为这将更直接和全面地影响全球未来的发展，这是一种基于共享经济理念的新经济形态。近年来出现的共享金融的表现形式就更多样化了，其中最为我们熟知的应该就是“众筹”，这是通过社交网站的传播，让小型创业家、艺术家、学生等个人或群体对公众展示他们的想法和创意，获取关注和支援，进而得到所需要的资金援助，通过互联网和共享思维来加速小规模的资金累积，促使许多原来不可能实现的想法变成现实。而虚拟货币如比特币、以太币、瑞波币的诞生也是非常值得关注的，虚拟货币这种基于去中心化的“账本”（Shared Ledger），是基于互联网架构及共享互利思维上的新经济形态，这都是过去想象不到的经济形态，因而当代人们的生活也必然伴随着共享经济的进程产生变化。

◀ 透过区块链技术运作的比特币，特色是去中心化、全球通用，任何人皆可参与。
Bitcoin operates through regional block chain technology, featuring decentralization, global adoption, and participation by anyone.

Sharing economy

Key words: sharing economy, crowd funding, social mechanism

Sharing economy, also called rental economy, is a new social action of sharing manpower and resource, which consist of the invention, manufacture, allocation, transaction and consumption among divergent people and organizations. Sharing economy, meanwhile, exemplifies that anyone charging and creating values at the usage right transaction related to the idle assets. The concept of sharing economy was originated from the thesis *Community Structure Collaborative Consumption: A Routine Activity Approach* published by Marcus Felson and Joe L. Spaeth, who explained the activity of carpooling. That research discovered that the idle assets can be more effectively relocated by means of the "sharing usage," which was the predecessors of the prevailing sharing industries such as Uber, Zipcar, Airbnb, etc.

Based on the features and goals of sharing economy, the British scholar Rachel Botsman and the legendary investor Jim Rogers categorized these into sharing consumption, sharing manufacture, sharing study and sharing finance. The context of sharing finance is comparatively intriguing, which is a new economy derived from sharing economy. The development of sharing finance has been diverse in recent years, from which the "crowd funding" given the attributes of internet and SNS displays achievements of and raise funds for start-up entrepreneurs, artist, students and individuals. For the booming growth of visual currencies including Bitcoin (BTC), Ether and Ripple Credit (XRP), the thriving advancements of visual coins featured with libertarianism, anarchism and shared ledger are the brand-new economy advocating the mutual benefits. These unprecedented new economies come with the innovations of sharing economy as well.

建筑的共享

关键词：共享永续、再生能源、代谢派、大数据

永续性是当代建筑一个不得不回应的议题，因为当代土地所扮演的主要角色已转换成生产食物的工具，而共享与永续已是维护生产工具的必要手段。在面对石化能源不可逆的绝境下只有两条路能走：其一是发展可再生的永续能源，建筑师必须开始思考，不再是“让当下保持不变”，而应致力于“创造可持续性的未来”，通过新科技、新工法、新材料，赋予建筑自我代谢及共享的能力以延续建筑的生命周期，形成全新的建筑价值——具有变动性的“相对永恒”，以适应瞬息万变的城市；第二则是建立共享机制来达到“节能”的目的，通过互联网和大数据等科技手段，产生如共享单车、共享办公室、共享书店、共享住宿等新形态商业模式，希望减少能耗或提升空间使用效率。因而，在能源研究与数字科技的帮助下，我们让建筑设计联结自然能源与共享经济资源，促成再生和共享，把一种理念口号进化为可积极落实的建筑指标。

共享思潮给当代建筑师最大的挑战与激发，不是概念的发明或原创，而是如何通过空间的手法与策略来创造共享的机会与建立共享的机制。不能只依赖过往单一线性的过程，而是必须打开多元化、跨领域的网络，在大数据的信息基础上，与工程师、科学家、环境专家、政府部门建立对话及合作关系。“数据共享”是当代建立跨界沟通的客观理性的语言，设计者必须具备能将一切信息化为数据的能力，同时通过精准的模拟物理环境运算，才能真正有效地反映在建筑设计上，赋予建筑自我代谢以及回应环境变化的能力，掌握永续发展的未来。

“代谢”一词源自于生物学，指生物体持续不断地与周遭物理环境进行能量交换与物质替换，借此完成自身更新来适应内部与外界的变化，是一个生命体通过自我更新以获得继续生存的过程。面对着第二次世界大战后几乎被摧毁的日本，丹下健三等建筑师将这一生物学的概念应用在建筑理论上，提出著名的《新陈代谢 1960，新都市主义的提案》：城市建筑设计也应该像生物的新陈代谢那样，在有机和共享机制的动态过程中成长。因此站在“相对永恒”的思考脉络上，就不难理解 20 世纪 60 年代日本丹下健三及黑川纪章等建筑师所发起的代谢派（Metabolism）的建筑行动和彼得·库克（Peter Cook）主导的“建筑电信”（Archigram）的建筑思潮。只可惜，在当时数字信息、绿能科技发展及大时代氛围不足的时空背景限制下，他们所采取的策略主要是从建筑构造上实验可替换、可移动的结构设计，不免仍有些乌托邦式的科幻神话。时至今日，科学家们在能源发展与数字科技上取得了日新月异的突破，我们有能力在自然界找寻潜在的秩序和现象，并将潜在的能量流（Energy Flow）整合收纳到建筑中，因而造就了共享、多元、复杂、精密、有机、包容的设计策略和建筑形态。在当代持续城市化的进程中置入永续发展的因子，带动建筑思潮向更具体的“共享永续”的目标前进，成为全球建筑师共同努力的方向。

Sharing architecture

Key words: Share sustainability, Renewable Energy, Metabolism, Big Data.

The issue of sustainability is of great importance, for the sharing and sustainability are crucial for land cultivations that nurture foods. There are two options given the development of depriving petrochemical energy. The first one is to boost recyclable energy. The architects, meanwhile, have to "step out of the comfort zone", and "pursue a sustainable future." The application of modern technology, crafts and materials spawns the capacity of metabolizing and sharing, which prolong the life cycle of building. Responding to the ever-changing city, a new architectural value is established that is volatile "relatively eternal." The second one is to constitute the sharing mechanism accomplishing "energy saving." The approaches using internet and Big Data stimulate the effective evolution of sharing bike (like OFO), sharing office (like Myspace), sharing book store, sharing accommodation, etc. With the advancements of energy research and digital technology, the architectural designs are consolidated with the recyclable natural energy and sharing economy. The originally political-correct expressions, thereafter, progresses into the pragmatic architecture indices.

Contributing rather than conceptual innovation or originality, the greatest challenge and inspiration from sharing ideas is to create sharing opportunities and regulations by means of space skills and strategies. Instead of depending solely on the linear-approach, the architects must take advantages of Big Data growth, and collaborate with engineers, scientists, environment experts and government. "Data sharing" is the most objective communication media. Designers need to cultivate the capability of digitalization, and to simulate the physical calculation of architecture design, which hatches the architectural capacities of self-metabolism and environmental reaction, and attains the sustainable future.

Derived from biology, the word "metabolism" is the set of life-sustaining physical and chemical transformation within the cells of organisms, which is adjustable to accommodate the environmental changes. Confronting the Japan's post world war II reconstruction, the several architects including Kenzo Tange stated in this renown "Metabolism 1960" that "buildings and cities should be designed in the same organic way that life grows and changes by repeating metabolism." Thinking based on the idea of "relatively eternal," people are aware of the 1960s Japanese architects Kenzo Tange and Kisho Kuroawa who proposed the architectural action of "Metabolism," and Peter Cook who dominated the architectural idea of "Archigram." Given the limited growth of digital information, green technology and constrained environment, the applicable strategies were mostly confined to the structural designs replaceable and portable, which were somewhat utopian. Today, scientists have been making huge progress on energy development and digital technology, which assist us in exploring the potential discipline and phenomenon. People consequently merge the divergent energy flows, and stimulate the design strategies and architectures that are sharing, diverse, complex, sophisticated, organic and inclusive. The implementation of sustainable factors during the progress of urbanization helps establish the prevalent architectural movements leading to the "sharing sustainability."

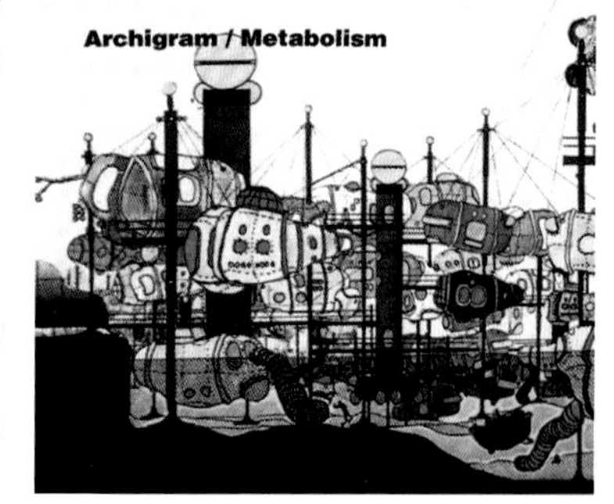

► 《建筑电信·代谢派》一书中，罗恩·赫伦（Ron Herron）的海边泡泡建筑，相互共享和依存的建筑想象。

In his book *Archigram Metabolist*, Ron Herron's seaside bubbles of architecture shown an architectural imagination of sharing and and depending on each other.

老板电器——厨源

位于上海的精华地段，设计师颠覆传统思维，将最重要的一楼空间解放出来，通过多元领域的共构碰撞，为戴维 · 谷曼（David Guzman，米其林三星主厨）、王强（工业设计师）、保罗 · 帕瓦（Poul Pava，丹麦艺术家）及江英俊（厨艺名家）打造了以美食生活场景为主轴的共享空间。

Robam – culinary

Robam, located on the prime location in Shanghai City, exhibits the unconventional style that the first floor adornments manifest diverse ambiences, of which the design collaborations among David Guzman (Michelin 3-sart chef), Qiang Wang (Industrial designer), Poul Pava (Danish artist), Ying-Jin Jiang (famous chef) and CHU-studio present to people a culinary sharing space.

◀ 老板电器“厨源”的共享理念，结合不同跨界品牌的共同运营。
Robam’ "kitchen source" sharing concept combined with the operation of different cross-border brands.

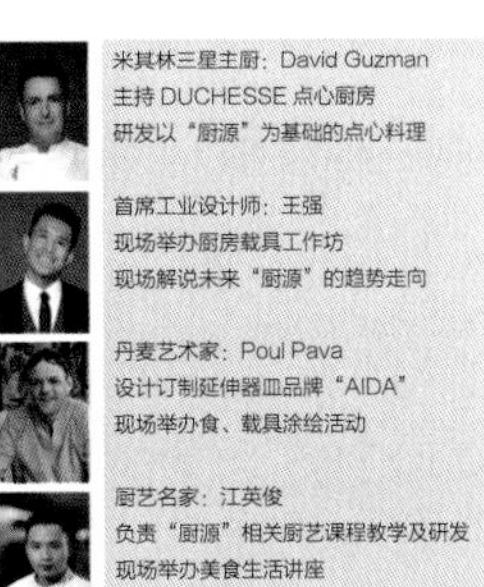

◀ 老板电器“厨源”的共享思维，各领域的跨界合作。
Robam’ "kitchen source" sharing concept, cross-border cooperation in various fields.

新零售的共享

关键词：去时化、在线一线下、体验店

建筑大师雷姆 · 库哈斯（Rem Koolhaas）曾提过商业行为（购物、新零售）将是未来人类重要的公共社交生活，而随着共享经济的崛起必然会建立及带动一种全新的商业模式，并重新建立人与人之间的互动关联。这也是笔者要特地谈及新零售共享的主要原因，因为这也是和我们生活基础需求联系最紧密的产业。

过往 1.0 时代的实体零售店，靠的是人力的单一服务，最典型的就是夫妻店（夫妻两人共同经营的小型商店）。后来进化到了以阿里巴巴为代表的新零售 2.0 时代，即所谓的 O2O（Online To Offline）时代，零售业开始迈入互联网时代，电商大量崛起，冲击并取代了过往单一线性发展的实体商店。而到了后电商时代，大数据、人工智能等科技也全面渗透进零售领域。随着电子商务的全面发展，其广告运营成本和竞争也日趋白热化，终于催生了新零售 3.0 时代的来临，即商家（企业）以 2.0 时代的互联网为基础，运用大数据、人工智能等技术手段，对商品的生产、流通与销售过程进行升级。同时 O2O 模式开始进化成全方位的去时化（time-out）发展，并对在线服务、线下体验进行深度的共享融合，更重视共享、体验及差异化的服务。并且实体店会以小规模的尺度遍布各地，形成实体的互联网络系统，相互支援及共享资源和服务。服务和体验成为零售最重要的核心价值，服务指的不是与商品相关的服务，也不是单指店面体验的具体服务，而是要提供各种类型更宽广的“生活服务”，并上下渗透到新时代中出现的两个重要族群：“不想动的年轻人”与“动不了的老年人”。因此，新零售的概念将从狭义的、单一的、线性的“商品买卖”，发展到广义的提供一切相关服务与活动，这是零售业结构性的改变，也是近年来阿里巴巴、小米、亚马逊等企业纷纷从线上跨入线下的主要原因。笔者设计的几个 3.0 时代的新零售空间，强调共享价值、重视体验感，最能体现时代特色。

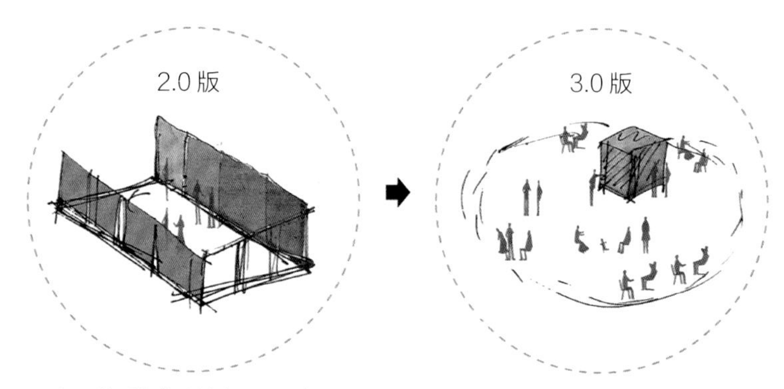

▲ 新零售时代的终端店更重视多元共享的社交理念。

Terminal stores in the new retail era pay more attention to the social concept of multiple sharing.

Sharing new retail:

Key words: time-out, Online to Offline (O2O), experience store

Rem Koolhaas, the top architect, mentioned once about that the business behaviors (shopping and new retail) will be the dominant public life in the future. The rising sharing economy will definitely constitute a new business model, and reconnect the relationship among people. This is the main reason that I discuss about the sharing new retail, which is intimately related to our daily life.

Retail 1.0 relies on manpower services, and the husband-and-wife shop managed by individual couples is a classic example. The so-called O2O (Online to Offline) business model mentioned by Alibaba Group Chairman Jack Ma, Retail 2.0 involves massively with the development of the internet and e-commerce that take over the traditional physical goods store. In the post e-commerce era, the advances of Big Data and AI has been comprehensively applied in the retail industry. The greater scale launching the e-commerce stores, the greater intensified advertisement costs and competitions, which provoke the birth of Retail 3.0. referred to as the stores or companies, operated based on the internet advances in Retail 2.0, that deploy the technological skills such as Big Data and AI upgrading the procedures of manufacture, distribution and sale. The O2O business model, meanwhile, has begun revolving into the complete time-out business pattern. The O2O starts merging the online service and experience, while focusing more on the services of sharing, experience and diversity meanwhile. The physical-goods stores, concurrently, are expanded widely on the small scale constituting into the physical internet system that share the resources and services as well. When services and experiences become the core value of retail business, service refers to neither the commodity service nor store experience, but the extensive "life service" .

There come two crucial groups in the new generation, which are the "vegetated youngsters" and "inconvenient elderly". Referring originally to the narrow, single and linear "goods transaction", the concept of new retail business develops extensively into the ultimate services and activities. This is the structural variation of retail industry that motivates Alibaba, Xiaomi and Amazon to expand from online to offline. I hereby present to you the space design works of Retail 3.0 strengthening that the sharing value concerning experiences signifies its treasure the most.

北京 MIBA 酿酒大师艺术馆

酒窖不再只是单纯的存酒空间，竹工凡木设计团队将摄影棚、酒吧、艺术馆、教室等不同的空间重新整合，通过不同文艺活动的相互交织，共同构成酒文化的交流场所，这里已成为北京重要的酒文化地标。

MIBA Beijing

In MIBA, the wine cellar represents more than a wine storage space, of which the CHU-studio synthesized the space characteristics of film studio, bar, art gallery, classroom, etc. that constituted into a famous wine culture space in Beijing.

◀ 北京 MIBA 酿酒大师艺术馆，将传统老厂房建筑改造成以共享酒文化为主题的生活社交场所。
At MIBA Wine Master Art Museum in Beijing, the traditional old wine factory building has been transformed into a social living place with the theme of sharing wine culture.

◀ 北京 MIBA 酿酒大师艺术馆酒吧，不浪费任何可能的机会，将老酒罐变成一个生活社交的场所。
At MIBA Wine Master Art Museum in Beijing, by seizing any chance, it really become a social living place.

托斯卡纳未来店

本案位于南昌，竹工凡木设计团队打破传统瓷砖店封闭、单一的展售形态，打造出以铺贴为主题的文艺生活的公共场所。

Toscana - Future store

The CHU-studio was responsible for the interior design work for Toscana, a tile shop situated in Nanchang City, that the design team created an art-and-cultural public area displaying the theme of tile facing.

▲ 新中源陶瓷的托斯卡纳未来店，破除了传统零售店的单向线性思维，追求生活艺术平台的搭建。
The Tuscany Future Store of Xin Zhongyuan Ceramics breaks away from the one-way linear thinking of traditional retail stores by pursuing construction of life art platform.

▲ 新中源陶瓷的托斯卡纳未来店，重视文化与生活，贩卖的不是单一的瓷砖，而是一种生活的方式。
The Tuscany Future Store of Xin Zhongyuan Ceramics stress culture and life, it is selling a way of life, not just tile.

沟通的共享

关键词：沟通、非预期成效、多元媒介

管理学家切斯特·巴纳德（Chester Irving Barnard）认为：沟通是把一个组织中的成员联系在一起，以便实现共同目标的手段。沟通是一个将思考、观念、情感、价值，传递给对方的连续动作，是一个复杂的过程，是人与人之间、人与群体之间思想与情感传递和反馈的过程。然而，若沟通的本质是一种历程（Process），其媒介（Media）就显得格外重要了。互联网时代，沟通的媒介非常多元化，除了实体的言语沟通外，透过科技手段和 app 的介入，如微信、抖音、微博等，我想还有太多的例子，这些数字媒介工具创造了更多交流的通道及形式，让我们沟通的可能性更加灵活而多元。但如今无论哪一种形式的沟通，都是架构在信息共享的方式上，我们已将生活的全部链接给科技，无论是生活还是工作，抑或是情感的传递都基于科技和网际网络所建构的数字媒介，通过大量信息的共享来达到目的。

举一个自身的例子，笔者本身在几所大学教书，同时也指导毕业设计的课程，随着内容的需要，我与同学们已共同建立起一种在数字媒介的平台上上设计课的机制，并鼓励同学们实地前往各种与研究主题相关的展览考察，通过直播和在线交流的方式来相互学习。同时，我也将不同系别的学生跨系组成讨论群，比如将室内设计系、建筑系、景观系的同学合并上课，增加多元碰撞的机会，同时共享设备（如快速成型及激光切割机），以达到知识及技术共享，以便创造出更出乎意料的学习效果。美国宾夕法尼亚大学建筑系老师罗兰·斯努克斯（Roland Snooks）也表示，传统的建筑教育是主观的从上而下的思维，而他强调当下更有效的学习方法，应该是由下而上的组织化学习。这也恰巧回应了德国斯图亚特大学阿基姆·门格斯（Achim Menges）教授认为当代更有效的学习方法，应将过去传统师徒制的设计课机制转换成为新形态的共同研究工作模式，许多学校也都已开始跟进这样的设计课理念。最有创造力的沟通是透过多元的行为主体，通过各种载体实现信息多向流动过程，而当代的开放思维及成熟技术载体正好提供给我们这样的机会。

◀ 由下而上的管理反馈系统已是近来企业管理的一门显学。
The top-down management feedback system has become a subject of enterprise management recently.

Sharing communication

Key words: communication, unexpected result, multi-media

Chester Irving Barnard, a management theorist, believed that "communication is the accomplishing approach to connect the group members." Communication is the constant actions expressing among people the thoughts, ideas and affections, which are a complex and reaction progress between people and communities. If the essence of communication is a type of process, its media is of great importance as well. Corresponding to the impact from the internet, the communication medias are divergent, encompassing verbal and technological channels of numerous apps such as Twitter, WhatsApp, WeChat, Line, FB, Instagram and Tik Tok. These diverse digital medias strengthen the communication effectiveness. The digital medias based on teehndogy and internet, has been nowadays comprehensively conveying massive information for our life, work and emotion.

Taking me as an example, I have been teaching and advising graduation design works in several colleges. To provide the appropriate course structures, the students and I collaborated on establishing the digital media platform over which lecturing classes. Meanwhile, I also encourage the students to participate and engage in the field researches and exhibitions. The students can learn and share information over the live broadcast and online platform. I suggested to initiate the interdisciplinary classes uniting the students from distinct departments, e.g. interior design, architecture and landscaping, that the students effectively and efficiently exchange knowledge and share equipment (RP and laser cutting machine). This type of learning environment is productive and educative. Roland Snooks, an architecture teacher in University of Pennsylvania, pointed out that the modern bottom-up learning is more enlightening than the traditional top-down. Thinking agreeably, the professor Achim Menges asserted that the traditional master-student relationship should be replaced by the new co-worker relationship, which has been accepted by many school such as Bartlett. The most innovative communication is based on the multiple behavioral identities carrying divergent information, and the contemporary thoughts and mature technologies have created the greatest opportunities for us.

卜天静　北京
ISENSE DESIGN 北京吾觉空间设计创始人

引 · 共享

■ 2017 年度媒体十大流行语

2017 年，“共享”一词被评为“2017 年度中国媒体十大流行语”，这个榜单基于国家语言资源监测语料库，利用语言信息处理技术，结合人工后期处理提取、筛选而获得。语料来源包括国内 15 家报纸 2017 年 1 月 1 日至 2017 年 11 月底的全部文本。这些报纸包含政府机关报、地方都市报和发行量较大的晚报，语料规模近 5 亿字次，代表了中国主流媒体关注点和语言特点及其变化指征。

“2017 年度中国媒体十大流行语”对于“共享”的定义是：“共享”是共享经济中的核心理念，强调物品的使用权而非所有权。共享经济是公众将闲置资源通过社会化平台与他人共享，进而获得收入的经济现象。2016 年，共享单车的兴起将共享概念带入了人们的视野。2017 年，共享经济更加发展壮大起来，涉及行业不断增加，规模不断扩大。共享单车、共享汽车、共享雨伞、共享充电宝……种种创新发挥着人们的想象力，同时也是对社会闲散资源进行合理利用的尝试。

尽管“共享”的概念火遍大江南北，但纵观各大主流媒体关于其含义的探讨，大多都侧重于经济理念、资源调配等物化层面。在飞速发展的时代进程中，人们更加关注事件的结果，而不去追寻源头，更有甚者只是单纯地随波逐流。我们常常忽略了关于共享精神层面的重要意义，例如正确价值观的共享、社会责任与使命感的共享、知识共享、爱与尊重的共享……

introduction-sharing

Not long ago, “Sharing” was entitled as one of the “2017 Top Ten Chinese Language Media Buzzwords,” of which the selection process encompassing the 500-million linguistic data signifies the majority concerns, linguistic habits and variation features by the Chinese mainstream media who defines the term “Sharing” as the followings.

“Sharing,” the core concept in the “sharing economy”, focuses on the right of goods usage rather than that of goods ownership. As the economic phenomenon, the “sharing economy” advocates that the sharing of idle resources over the socialized platform generate incomes. In 2016, the emerging bicycle-sharing services brought the idea of “sharing economy” to the public. In 2017, the scale of “sharing economy” was boosting. Afterwards, the developments of bike-sharing, car-sharing, umbrella-sharing, charger-sharing, etc. exemplify the people’s innovations and attempts to logically utilize the idle resources as well.

Though the comprehensive popularity of the concept “Sharing”, most of the mainstream media, however, discuss it from the materialized perspectives of economic ideas, resource allocations, etc. living in the fast-growing era, people care more about endings than about beginnings, the worst, nevertheless, is following with no plans. Regularly, we overlook the importance of spiritual meaning of “Sharing,” such as its value concepts, social responsibilities, missions, knowledge, love and respects.

▲美国漫威漫画公司历年票房最好的作品

Marvel Comics's the bast box office work over the years.

卜天静　北京
ISENSE DESIGN 北京吾觉空间设计创始人

平凡世界中的超级英雄

Superheroes in real world

■最终的胜利，都是价值观的胜利

The ultimate victories, the value concept victories

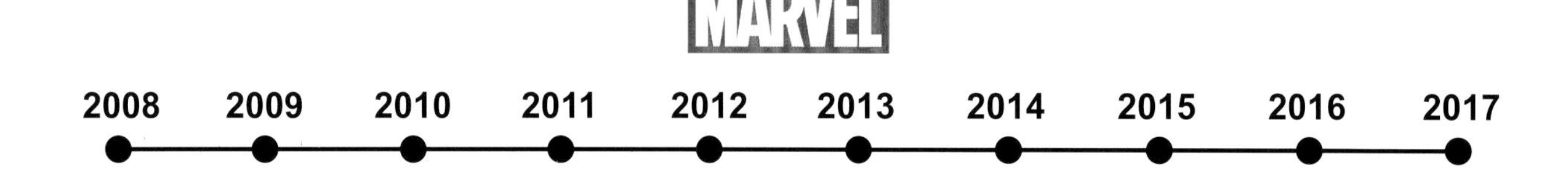

2008 年《钢铁侠》（*Iron Man*）在银幕亮相，截至 2017 年已是漫威电影宇宙开启的第九个年头。随着《雷神 3》的上映，9 年来 17 部漫威电影全球票房超过 130 亿美元。除了成功的商业运作以及 IP 的养成重塑外，其题材提倡的价值观也是吸引观众的原因之一。这些故事抓住了大多数人，尤其是男性观众与生俱来的英雄情结——光明战胜黑暗，爱与正义的胜利，这是跨越国界的普世价值观。各路“英雄”与“超人”，有的生而拥有超能力，有的精通科学，“但最终的胜利，都是价值观的胜利（此处引用前广州设计周总干事张宏毅先生生前常讲的金句）”。

It has been 9 years since the movie *Iron Man* released in 2008 by Marvel Cinematic Universe. Following the movie *Thor: Ragnarok*released in 2017, the accumulated box offices of the 17 Marvel Comics movies over 9 years has been more than 13 billion dollars. In addition to successful business model and top IPs (Intellectual Property), the value concepts of these movie topics are one of the reasons enticing people. Among the alluring factors, especially the natural-born male heroism – light defeats darkness, love and justice – are the universe value concepts beyond the borders. The assembly of “heroes” and “superman” that is imaginative merges the super power and science signifies the value concept victories (popular quotes by Mr. Zhang Hongyi, the Director General of Guangzhou Design).

■一个演员能否改变一个国家?

Can an actor change a country?

2017 年，一部电影《摔跤吧，爸爸》赚足了观众的眼泪和掌声，影片中的主角扮演者，是印度影帝阿米尔 · 汗（Aamir Khan）。他的电影大都带有强烈的社会意义：《三傻大闹宝莱坞》和《地球上的星星》题材是教育问题，《摔跤吧！爸爸》直指性别歧视。

2012 年至 2014 年，阿米尔·汗主持电视节目《真相访谈》，他和他的团队走访全国，调查取证，尖锐揭露印度最敏感的社会问题——性侵、堕胎、包办婚姻、家庭暴力、种姓制度等。他说：“我演电影时，在不同的角色中，体验过不同的人生。还有另一种人生，就是我自己的人生：卸去演员的身份，作为一个人，以我的方式存在……因为我也是这个社会中的一分子。”

2012 年 9 月，阿米尔·汗登上《时代周刊》亚洲版的封面，被选入“全球百位最具影响力人物”。评论写道：“他直面印度的社会弊病，打破了宝莱坞的固有模式。一个演员能够改变一个国家吗？” 阿米尔 · 汗接受媒体采访时说：“我确实相信电影有推动改革的潜力，当然电视也一样……我们讲述的故事、刻画的英雄人物，可以激励民众，给人们带来希望，触动人心。解放人性中善良的一面，让它茁壮成长，这是一个创作者可以对世界做出的贡献。这是一种选择，不是我们的义务，但是如果我们想这样，是可以做到的。”

媒体评价阿米尔 · 汗在戏里戏外都是改变一个国家的真“英雄”。作为演员，他用尽全力扮演好每一个角色，用作品传递现实存在的光明与黑暗。戏剧之外，坚定不移地面对一个国家的弊病与缺漏，与民众共享正义的存在与勇气。

以上故事或许已经耳熟能详，在此分享，主要是想要分享几个观点：

第一，共享的行为并不一定以物质为媒介，也不一定以可量化的经济价值来衡量。共享关注、共享信念、共享希望、共享勇气、共享责任感……这些都是弥足珍贵的精神层面的共享。

第二，共享的方式是多元化的。每个人都扮演着不同的社会角色，在这个越来越开放和多元化的时代，只要拥有正确的价值观和信念，无论任何国籍、种族、行业、性别，都可以用自己的方式去影响世界。

提起“公益”，很多人觉得离自己很遥远，但事实并非如此。“公益”是一种理想和价值观，它不等同于豪迈挥金的资助，或者镁光灯下的作秀。作为一名青年设计人，我们至少可以从自身出发，呼吁或者解决身边的一些小问题，以设计之力点亮周遭，相信更多志同道合的力量加入之时，星星之火也足以燎原。只有体验过才知道，利他与利己所获得的成就和价值感是不可相提并论的。

In 2017, the big success movie *Dangal* won the audiences over. Aamir Khan, the movies' protagonist and India's famous actor, showed his social concerns in his films, of which the *3 idiots* and *Taare Zameen Par* present the educational issues, and the *Dangal* pointed out the issue of gender discriminations.

From 2012 to 2014, Aamir Khan hosted the TV show *Satyameva Jayate*, an Indian language of "Truth Alone Prevails," that comprehensively conducted numerous investigations and thereafter bluntly unveiled the most sensitive social issues in India such as sexual assault, marriage arrangement, domestic violence, and caste system. Aamir said that "When I was in a movie showing different character and life style, there was another me, who is real-life human as an indivisible part of this society ".

In September 2019, Aamir was on the cover of Asia Times and chosen as the "100 most influential people in the world." The report commented that "He faced the social issues in India, and breached out the Bollywood intrinsic modes. Can an actor change a country?"

▲阿米尔·汗登上美国《时代周刊》封面

阿米尔·汗，印度著名电影演员、导演、制作人和主持人，他主演过许多部卖座的印度佳片，被美国《时代周刊》评选为“印度的良心”。

Amir Khan appeared on the cover of Time Magazine

Amir Khan, a famous film actor, director, producer and presenter from India, has starred in many popular Indian films, and was voted the conscience of India by the US Time.

Aamir had said in an interview that "I do believe that movies can make changes, so can TVs... That the stories we tell and the heroes we create is inspiring, encouraging and touching. Unleashing the positive human nature and flourishing it, which are feasible contribution a creator makes. It is an option, rather than an obligation. But we can make it happen if we want."

Being reported as a "real hero," Aamir Khan does his best to act in the film, convey the corporeal light and darkness. He, concurrently in real life, highlights resolutely the nation's corrupt customs and ignorance, fearlessly "sharing" the justice and courage with people.

The two stories, mentioned above, are well-known. I, accordingly, want to share few perspectives, of which, firstly, "sharing" does not have to be either physical or quantifiable. It can be sharing attention, belief, hope, courage, responsibility, etc. that are spiritual and cherished "sharing." Secondly, "sharing" can be diverse. Everyone is unique, and each of us having authentic value concept and belief can make a change to the expanded and divergent world, regardless of nationalities, races, careers or genders.

"Public welfare" is not as distant as people imagine, which represents the mission and value concept that is unlike the colossal fundraising or fashion show. As a young designer, I advocate that we start by examining and solving possessed problems. Calling for more presence of like-minded friends, the power of design can be influential. By practicing what one preaches, everyone will realize that altruism is more fulfilling and valuable than is egoism.

■设计，是设计人的“超能力”

Design, the super power of designers

前文谈到社会责任感、公益心以及共享精神。纵观世界，早已有许多设计人、建筑师怀着社会责任感和使命感，作为先行者不断践行着，如果说纸笔是鲁迅先生唤醒民族，与腐朽抗争的武器，那么设计，就是我们设计人的“超能力”。

The social responsibility, public-welfare awareness and sharing spirit, mentioned in the prior sections, have been put into practice by many pioneers such as designers and architects. If papers and pens are the weapons for Mr. Lu Xun to protect people and fight against corruptions, design is the “super power” of designers.

▲普利兹克建筑奖历届获奖者

The pritzker Architecture Prize

■普利兹克建筑奖——现代建筑史的缩影

Pritzker Architecture Prize - the epitome of the modern architecture history

普利兹克建筑奖被誉为建筑界的诺贝尔奖。

以下是普利兹克奖的官方描述：表彰一位或多位当代建筑师在作品中所表现出的才智、想象力和责任感等优秀品质，以及他们通过建筑艺术对人文科学和建筑环境所做出的持久而杰出的贡献。

这一国际性奖项由美国芝加哥普利兹克家族通过旗下凯悦基金会于 1979 年创立，每年评选一次，授予一位或多位做出杰出贡献的在世建筑师。该奖项通常被誉为“建筑界的诺贝尔奖”和“建筑界最高荣誉”。

奖赏包括 10 万美元奖金和一枚铜质奖章，授予一位或多位获奖人，颁奖仪式选择在世界各地的著名建筑物内举行。

To honor a living architect or architects whose built work demonstrates a combination of those qualities of talent, vision, and commitment, which has produced consistent and significant contributions to humanity and the built environment through the art of architecture.

The international prize, which is awarded each year to a living architect/s for significant achievement, was established by the Pritzker family of Chicago through their Hyatt Foundation in 1979. It is granted annually and is often referred to as “architecture’s Nobel” and “the profession’s highest honor.”

The award consists of $100,000 (US) and a bronze medallion. The award is conferred on the laureate/s at a ceremony held at an architecturally significant site throughout the world.

▲普利兹克建筑奖章是根据路易斯・沙利文的设计而铸造的。路易斯・沙利文是芝加哥著名的建筑师和公认的摩天大厦之父。奖章的一面是奖项的名称，另一面则刻有三个词：坚固、价值和愉悦，呼应古罗马建筑师维特鲁威提出的三条基本原则：坚固、实用和美观。

The Pritzker Architectural Award is based on the design of Louis Sullivan, a steadfast Chicago architect who has been recognized as the father of skyscrapers. On the one side of the medal is the name of the award, while on the other side engraved three words: firmness, value and pleasure, which is echoing the three basic principles put forward by Rome architect Vitrurius: solid, practical and beautiful.

◄普利兹克奖 2017 年得主
Rafael Aranda、Carme Pigem 和 Ramon Vilalta
图片来源：Javier Lorenzo Domínguez

▼普利兹克奖 2017 年得主作品
Les Cols 餐厅 Marquee. (2011) Olot，Girona，Spain. Hisao Suzuki.
图片来源：Pritzker Architecture Prize

▼普利兹克奖 2013 年得主伊东丰雄（Toyo Ito）作品
Meiso no Mori Municipal Funeral Hall 2006
图片来源：Pritzker Architecture Prize

普利兹克奖是到目前为止，在建筑界最具权威性的奖项，每一位获奖建筑师，都在特定的时期和背景下引领着建筑创作的潮流。

早期的获奖者贝聿铭、丹下健三等对现代主义建筑进行研究并促成国际主义建筑风格的多元化发展；安藤忠雄借用了现代主义的形式，对整个现代主义进行批判性改造；雷姆 · 库哈斯更注重建筑的社会性，要求建筑应对每种社会新问题做出回应；扎哈 · 哈迪德是普利兹克奖设立 26 年以来第一位获奖的女建筑师，她独特的创作方式，让大胆梦幻的数字建筑震撼世界，并多方位进行设计实践，涉足交通工具、产品、家具、时尚用品等几乎所有设计门类……

有趣的是，从普利兹克首届获奖者菲利普 · 约翰逊开始，历届获奖者并非全部是建筑专业出身，他们当中有许多来自其他领域，并且涉足多方位领域的研究。组委会的评判标准也从建筑的专业理论、个性风格，转而更多地关注建筑师对解决社会问题的参与及社会责任感的表现。

2012 建筑师王澍成为首位获得普利兹克奖的中国建筑师，组委会赞赏他对于传统文化回归与传承的责任感。伊东丰雄注重建筑与社会、人、环境的关系。坂茂潜心研究、开发可再生建筑材料，减少浪费，并常年投身于公益事业，以建筑创作践行人道主义援助。弗雷 · 奥托、亚历杭德罗 · 阿拉维纳在智利最北部的伊基克的金塔蒙罗伊社会住宅项目中，提出了“半成品住宅”的创意，充分体现了建筑师在设计中强调公民参与的重要性。

这些建筑师的实践有几个共性：

一、探讨建筑的动态观。自古以西方为代表，一直在追求永恒（与宗教文化有一定关系）。而可以灵活拆装的建筑，或在灾难时可以马上利用的结构，很长时间内满足了公众对于“临时性”场所的要求，并满足当代快节奏生活的需求。

二、关注普通大众的生活与文化根基。通过共享空间，促进人和人之间的沟通，摸索“新”与“旧”的重构组合。寻找文化历史传承与适应当下生活标准的平衡点，也是全球性探讨的议题。

三、注重团队协作与公众的参与。建筑师开始注重使用者在设计中的参与度，通过共同协作完成建筑的设计或建造，可以增加使用者的归属感与责任感，使之与空间产生更亲密的情感。

从建筑可以窥见人类文明的进程，而普利兹克建筑奖则是现代建筑史的缩影。

诚然，每个建筑师、设计人都有属于自己的追求和信条，我们并不能用道德绑架所有设计人都要去关注社会，然而就如前文阿米尔 · 汗说的话：“这是一种选择，不是我们的义务，但是如果我们想这样，是可以做到的。”设计是设计人生存的手段，也是我们拥有的、影响世界的力量。

Pritzker Architecture Prize, by far the most authoritative prize in architecture, has been awarding numerous noted architects representing specific time and creation.

The earlier prize winners, including Ieoh Ming Pei and Kenzo Tange, exemplify the modernism of architecture that motivated the diversified development of nationalism architectural style. Applying the modernism, Tadao Ando modified and reestablished the modernism. Rem Koolhaas, concerning the social mechanism of architecture, requested that architectures respond to every new social issue. Zaha Hadid, the first awarded female architeet in 26 years since prize development, uniquely present design works amazing the world. Meanwhile, she designed diversely, incorporating transportation vehicles, products, furniture, fashions, etc. The fun fact, however, is that for those prize winners since Philip Johnson, they have been prestigious in distinct expertise other than architecture. The jury members have been constantly modifying the concerns, ranging from professional architectural theories and genuine styles, to social issues resolutions, public participations and social responsibilities.

In 2012, Wang Shu was the first Chinese architect winning the Pritzker Architecture Prize, of which the jury members admired his affections to traditional cultures and responsibilities to heritages. Toyo Ito concentrated on relationship among architecture, people and environment. Shigeru Ban dedicated to developing recyclable materials and minimizing wastes. He was also active in public welfare activities, and showed his support to humanism. Alejandro Aravena, the Chilean architect, won 2016 Pritzker Architecture Prize presenting the "half-finished homes" residential projects at Quinta Monroy in Iquique, the northern Chile. His inventive idea exceptionally explained the importance of civil participation in design.

These architects have something in common as following:

A. Exploring the dynamic architectural perspectives. The western worlds have been always symbolizing the dedication pursuing eternity, which is related with religious culture. Creating the architecture easy to assemble and dismantle, especially flexibly-accessible for urgency, the designers effectively solved the needs for either "provisional" living space or efficiency-requirement lifestyle.

B. Contemplating the people's life and cultural foundation. The idea advocating "space sharing" have been facilitating people's communications, the old-and-new consolidation, discovering the heritage of culture and history, and accommodating to global standard of balanced life.

C. Minding the team work and public participation. The architects started noticing the owners' participation in design works. The collaborative attitude strengthens the owners' sense of community and responsibility that intensify the intimacy between owners and spaces.

Architectures depict the progress of human civilization, and the Pritzker Architecture Prize represents the epitome of the modern architecture history. Each architect has different goals and missions, which are not mandatory to be ethical or social-concern. Recalling from Aamir Khan who claimed that "It is an option, rather than an obligation. But we can make it happen if we want," we have to know, however, that design signifies not only the human's strategic survival, but also the power to change the world.

坂茂 Shigeru Ban

■抚慰心灵的建筑才是永恒

The soothing architecture lasts forever

坂茂，2014 年普里兹克奖获得者。

坂茂坦言：“财富与权势是无形的，所以他们雇佣我们建造宏伟的建筑，来彰显他们的财富……我为无法服务于大众而感到失落，尽管有那么多人，因为自然灾害流离失所。”多年来坂茂不断活跃在第一线，为灾后无家可归的人寻找快速、经济、可持续的方案，用建筑帮助那些需要的人。

1986 年，早在“可持续性”成为一个流行词之前，坂茂已开始实验利用纸管作为建筑材料。没有任何先例可以参考，新材料的开发过程困难而漫长。好在结果令人欣喜，纸的硬度、防水、防火问题都可以解决。但研究只是第一步，即使试验成功了，他的专利却“不获得批准，没有人会相信可以用纸来建房子”。1995 年，坂茂在富士山的山中湖旁用 110 根纸管造了一幢度假屋，只是为了证明纸管可以被用于建造建筑。就在这一年，他的纸管结构开发获得了日本建设大臣颁发的永久性建筑认证。2000 年，坂茂受邀设计在德国举办的世博会上的日本馆，为了避免工业浪费，坂茂与德国建筑师弗雷・奥拖（Frei Otto）合作，用 440 根直径 12.5 厘米的纸管呈网状交织，表面再覆以纸模纸管构筑了一个巨大的网络薄壳结构。世博会闭幕后，这个薄壳建筑的所有材料都被运回了日本，制成了小学生的练习册。之后的几年中，坂茂带领学生又相继完成了许多受欢迎的建筑。但与此同时，他却对自己的职业生涯感到失望，作为建筑师，每天忙着服务于权贵，自己并没有服务于更需要帮助的大众。

1994 年，非洲南部的卢旺达两个种族部落发生了大规模的残杀。政府捐助的临时居所只有一张塑料席，因此当地 200 多万难民开始砍树自制帐篷，演变成严重的森林采伐，于是坂茂提议利用纸筒建造经济又坚固的庇护所。

1995 年，日本神户发生大地震，7000 多人遇难，包括教堂在内的建筑物全部倒塌。坂茂向当地神父建议用纸建造一座临时教堂慰藉那些信徒的心灵，但遭到了拒绝。之后，坂茂带领学生募款，联合麒麟啤酒公司赞助啤酒箱作为地基，建造了 50 多个临时庇护所。终于，坂茂得到了神父的信任，花了五周时间用纸筒建造了这座临时教堂。这座建筑原定计划是保存三年，但因为人

▼ 1986 年 坂茂开始研究纸管等可再生建材，并举办第一个展览。
In 1996, Shigeru Ban began to study renewable building materials and held the first exhibition.

们的喜爱，它矗立了整整十年。而后中国台湾发生了地震，坂茂团队提议捐赠这座教堂，于是拆掉了结构，在台湾重建，至今，这座教堂在台湾成为永久性建筑。

“Shigeru Ban, the 2014 laureate of Pritzker Architeature Priza, reflects this spirit of the prize to the fullest,” in the jury members' words, who admired him for his excellence in built work and for significant and consistent contribution to humanity.

“So I was very disappointed that we are not working for society, even though there are so many people who lost their houses by natural disasters. We are not working for society, but we are working for privileged people. They have money and power. Those are invisible. So they hire us to visualize their power and money by making monumental architecture.” - Shigeru Ban. Shigeru Ban has been for years finding for the homeless people the fast, economical and sustainable architectural solutions.

In 1986, much longer before people start talking about “sustainability,” Shigeru Ban begun testing paper tubes as building structures.

▼ 1994 年坂茂设计的为卢旺达的难民们提供临时庇护所的的纸筒结构
The paper tube structure designed by Shigeru Ban in 1994, provided temporary shelter for Rwandese refugees

There was no precedent to follow, and development of new materials was much harder. The testing result, however, was better than he expected. It was harder, waterproof and fireproof. Though the test was successful, Shigeru Ban's application for patent "was not granted, there was no one believed that the paper-house is viable." In 1995, Shigeru Ban built a resort by using 110 paper tubes by the side of Lake Yamanaka in Mount Fuji. " In fact, I don't have weekends or vacations," said Shigeru Ban, who wanted to prove that paper tubes are applicable for building structures. In the same year, his paper-tube structure was awarded the permanent building certification by Minstry of Construction, Japan. In 2000, Shigeru Ban was invited to design the Japan Pavilion in the World Expo hosted by Germany. To minimize the industrial wastes, Shigeru Ban worked with the German architect Frei Otto. They built a membrance structure consisting of 440 paper tubes, 12.5 cm in diameter and up to 40 m long. The roof skin comprises a specially impregnated textile and paper fabric. After the closing of the World Expo, all the construction materials were shipped back to Japan and were turned into the workbooks for the elementary-school students. Though Shigeru Ban led the students in accomplishing many popular architecture in the following years, he was disappointed at his profession as an architect. Rather than working for society, he was busy working for priledged people.

In 1994, there was a big disaster in Rwanda, Southern Africa, where two tribes fought

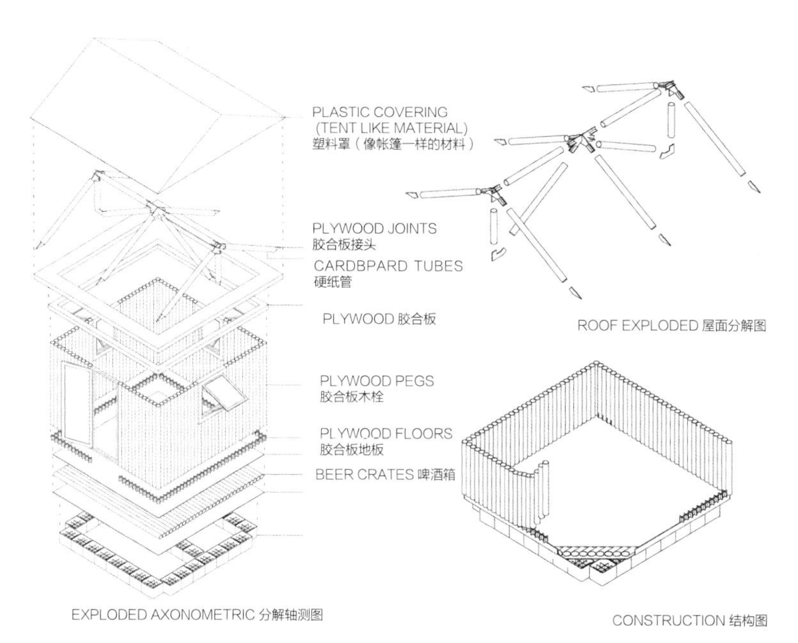

▲ 1995 年坂茂带领学生在神户建造 50 座临时庇护所
In 1995, Shigeru Ban built 50 temporary shelters with his students in Kobe, Japan

severely against each other and turned into a massacre. Over two million people became refugees. The refugee camp, organized by the government, had only one plastic sheet. Given the devastating environmental conditions, the refugees had no choice but to massively cut the trees, which became the heavy deforestation. Shigeru Ban, therefore, proposed to build the stronger shelters with paper tubes, which are much cheaper.

In 1995 when the Great Hanshin earthquake took place in Japan, there were more than 7,000 people died then. A large number of buildings, including the churches, had turned into ashes. Shigeru Ban proposed to the priest to build a church by using papers. The priest, nonetheless, refused to accept Shigeru Ban's idea as consolation. After that, Shigeru Ban and his students raised, jointly supported from Kirin beer company offering beer crates, and they built over 50 units of the shelters. They spent five weeks rebuilding the church. It was meant to remain intact for three years, but it stayed there for ten years because people loved it. Afterwards when a big earthquake occurred in Taiwan, China, Shigeru Ban proposed to donate the church, which was dismantled and sent over to rebuild in Taiwan, China. The building stayed, till now, as the permanent church.

▼ 1995年 神户纸教堂
Paper church in Kobe, Japan, in 1995

◀▼ 中国台湾发生大地震，坂茂提议捐赠日本神户纸教堂，教堂结构被拆除后运送到台湾重建，被留存至今，成为永久性建筑。
After a major earthquake in Taiwan, China, Shigeru Ban proposed to donate the Paper Church in Kobe. The structure of the church was demolished and moved to Taiwan for reconstruction. It has been preserved as a permanent building ever since.

至今，坂茂不断带领团队出现在全世界各个灾区，2000 年为土耳其地震灾民搭建了纸木宅；2001 年在印度地震后搭建了纸木宅；2008 年中国汶川地震之后，坂茂和来自日本及中国的志愿者用 33 天时间，为华林小学的 400 名学生建造了临时校舍，三座共 9 间教室；2011 年，中国四川雅安发生了地震，坂茂在这里用纸管结构设计了一所幼儿园。同年日本地震和海啸灾难过后，坂茂为福岛灾民设计了临时安置所。

建筑界也有人曾对普利兹克奖的评审标准提出异议，他们认为建筑师的社会关怀成了一种表演。而 2016 年，威尼斯建筑双年展的主题是“来自前线的报告”，建筑精英们看起来变得“不务正业”，他们跨出传统建筑学范畴的问题，讨论贫穷、灾难、污染、拥堵、隔离……坂茂不曾正面回应过这场热闹的讨论，称 “每个人都有权发表自己的想法”。他仍会在灾难发生时赶往现场。“没有一种所谓的‘模式’存在，可以套用到不同灾区。每一个地区的语境都不同，自然环境、建筑空间、人们的需求和地方文化……只能去到现场。”在所有人都在找寻模式提高效率与回报的当下，作为一代建筑大师仍奔波于各个灾难现场，亲自动手去创造符合当地人文状态的建筑，坂茂的执着和坚持令人钦佩与感动。

Shigeru Ban never stopped helping those suffering from natural disasters. In 2000 when the big earthquake occurred in Turkey, he built shelters out of papers. Likewise, in 2001 when earthquake shocked in India, Shigeru Ban showed his zealous for building up shelters as well. After the Wenchuan earthquake in 2008, Shigeru Ban and volunteers, coming from China and Japan, spent 33 days constructing the temporary school buildings for 400 students from Hualin Elementary School. There came the nine classrooms in three. In 2011 when an earthquake broke out in Yaan in Sichuan Province, Shigeru Ban built a kindergarten in paper-tube structures. In the same year after the earthquake and tsunami exploded in Japan, Shigeru Ban designed the temporary resettlement for the victims of Fukushima.

The construction industry once filed an opposing argument against the evaluation standards of the Pritzker Architecture Prize, whom being considered as “politically correct” show. The Venice Biennale announced the theme in 2016 as “Reporting from the Front,” which had made the elite architects seemingly being “unprofessional.” They discussed about poverty, disaster, pollution, congestion, isolation, etc. Shigeru Ban did not respond officially, and claimed that “everyone can express their own ideas.” He always hurried at once to the disaster scene. “There was no such a thing as a “pattern” applicable to different disaster areas. The context in each district is distinct, and people can only experience the natural environment, architectural space, people’s need and local culture when being at the scene personally.

坂茂认为：一座建筑，如果只是为了商业目的，即便是混凝土建造的，也是非常临时的；一座建筑，如果能够受到人们的喜爱，抚慰人们的心灵，即便它是纸管建筑，也可以是永恒的。

作为建筑师，坂茂无疑是大胆探索可持续性低碳建筑的先行者，更是用建筑救助心灵的革命家。他不但在危难时刻“共享”了自己的专业技术，更“共享”作为设计人的社会责任感及信念。在笔者看来，他的纸筒和建筑就是属于他的超能力，当之无愧为设计人的楷模与当代的英雄！

“A concrete building, if for commercial purpose, can be transitory. A paper building, if being loved by people and showing consolation, can be eternal,” by Shigeru Ban.

As an architect, Shigeru Ban was no doubt the pioneer of sustainable low-carbon buildings. He is also revolutionary in rescuing minds. Not only did he “share” with people his expertise, also he “shared” social responsibility and belief as a designer. From my point of view, his paper tubes and buildings are genuine his super power. There is no reservation that Shigeru Ban was the role model and hero for designers.

▲ 坂茂亲临灾难现场指导纸建筑的搭建
Shigeru Ban guide the construction of the paper architecture in disaster scene

迈克尔·墨菲 Michael Murphy

■伟大的建筑给人希望

Great architecture can give us hope

迈克尔·墨菲，美国建筑师，毕业于哈佛大学建筑系，与合伙人艾伦·里克斯（Alan Ricks）于 2010 年创立了 MASS 建筑事务所，同时在哈佛任教。这两个不到 40 岁的年轻建筑师已经带领着设计、研究团队走过了三大洲 10 个国家，帮助了数以万计的人。MASS 建筑事务所的作品涉及设计、研究、宣传和培训领域，致力于了解建筑设计中各个层面上的决策所产生的短期和长期的连锁效应——对于居民、客户、社区和社会。MASS 建筑事务所的实践侧重于探讨建筑与健康和行为的关系，以及设计人类的成长、尊严、健康所需的人文和物质系统。

Michael Murphy, an American architect, graduated from the Department of Architecture in Harvard University, and found with his co-founder Alan Rick the MASS Design Group in 2010. He also taught at Harvard University. These two young designers, less than 40, have led the design and research teams to visit more than 10 countries and 3 continentals worldwide, and helped out thousands of people. MASS architectural works covered design, research, advocacy and training, and MASS had been dedicated to understanding the short-term and long-term chain reactions of architectural design decision – for the residents, clients, communities and societies. MASS focused on exploring the relationship among buildings, health and behavior. It designed the humanity and material systems required for humans' development, dignity and health.

◄ MASS 建筑事务所创始人 艾伦·里克斯、迈克尔·墨菲
图片来源： Simon Simard
Founders of MASS Architecture Firm: Alan Ricks, Michael Murphy
Photo source: Simon Simard

“建筑从来不是中性的，它要么可以治愈，要么会造成伤害。我们的任务是去研究、建造和倡导可以促进正义和人性尊严的建筑。”

—— Mass 建筑事务所

“Architecture is never neutral. It either heals or hurts. Our mission is to research, build, and advocate for architecture that promotes justice and human dignity.”

— Mass Design Group

▲ MASS 建筑事务所官网
Official Web-site of Mass

我们常说建筑、空间应该是有温度、有情感的。事实上人与“房子”的互动，无论在心理还是生理范畴都有着微妙的反应。每周六，父亲穿着旧 T 恤粉刷老房子的身影一直伴随着迈克尔 · 墨菲的记忆。直到某一天，迈克尔一家得知父亲腹腔中有一个肿瘤，生命只剩下三周时间。在等待死神降临的时间里，迈克尔卷起袖子开始接替父亲，修整房子。三周时间过去，噩耗并没有来临。三个月后，父亲加入到了迈克尔的行列，父子俩重新粉刷室内、修缮窗户，并重建阳台和外立面。时光流逝，迈克尔的父亲奇迹般地痊愈了。当他们站在房子面前欣赏劳动成果的时候，父亲对他说：“这座房子救了我的命。”第二年，迈克尔考入建筑学院正式开始学习建筑设计。

“伟大的建筑一定要那样标新立异吗？它们看起来太过罕见，并且只有小部分人可以享受到。”带着疑问和彷徨，迈克尔在保罗 · 法尔莫（Paul Farmer）医生的一场演讲中得到启发。保罗 · 法尔莫是一位长期致力于全球减贫领域的健康活动领袖，他在演讲中质问建筑师对于人与建筑之间健康关系的缺乏思考和忽视。在贫穷的地方，建筑反而是造成流行病传播的主要原因之一。差劲的通风、缺乏细菌控制的空间，会让一个单纯骨折的病人感染多种致命传染病病菌并因此而死亡。一个医院竟然给人带来更多病痛，“建筑师在哪里？” 保罗医生说。（其实早前坂茂也曾在演讲中质疑过地震时造成死亡的真正原因。他认为地震本身并不是主要的原因，事实上大部分人死于建筑的倒塌，建筑师有义务更负责任地去设计建筑，并提供经济、可快速实现的临时住所。）之后的暑期，迈克尔追随保罗医生来到非洲南部的卢旺达，住在布塔洛布雷拉地区的一栋房子里，这里曾经是卢旺达种族屠杀的监狱。（前文提到的 1994 年坂茂第一次将纸结构投放到公益救助中就是在卢旺达。）

We often mentioned that the architectures and spaces should be warm and emotional. The interaction between human and houses, in fact, has psychological and physiological effects. Michael Murphy always remembered that his father wearing a T-shirt was painting the old house every Saturday. One day when the family learned that the father had a tumor in abdominal cavity, and he had less than 3 weeks of life. Three weeks passed over and his father remained alive. Three months later, his father and Michael started painting the house, and overhauling the windows, balcony and exterior. Miraculously, Michael's father recovered from tumor. When they stood in front of the house and looked what had accomplished, his father said “this house saved me.” The next year, Michael was admitted to the college and begun studying architecture.

“Do great buildings have to be unconventional?” Michael wondered and questioned. He was inspired from Paul Farmer who was giving a speech. Paul Farmer, a noted leader of health issues in global poverty, argued during his speech that architects were negligent to the health relationship between people and buildings. Paul Farmer believed that the buildings were the main cause for epidemic spread. The terrible ventilated and lack-of-bacteria-control space would be contagious and lethal to a broken-bone patient. It can be hardly seen that a hospital brings pains to people. “Where are the architects?” said Dr. Farmer (in fact Shigeru Ban once earlier questioned during a speech the real cause for more earthquake deaths. Shigeru Ban believed that the building collapses, rather than earthquake itself, killed more people. This was why architects should responsibly design the stable buildings, and create the economical and fast shelters.) In the following summer breaks, Michael followed Paul Farmer to Rwanda, Southem Africa. They lived in a house in Butaro Burera District, originally a prison for genocide in Rwanda (where mentioned in the earlier section Shigeru Ban built the paper shelters in 1994.)

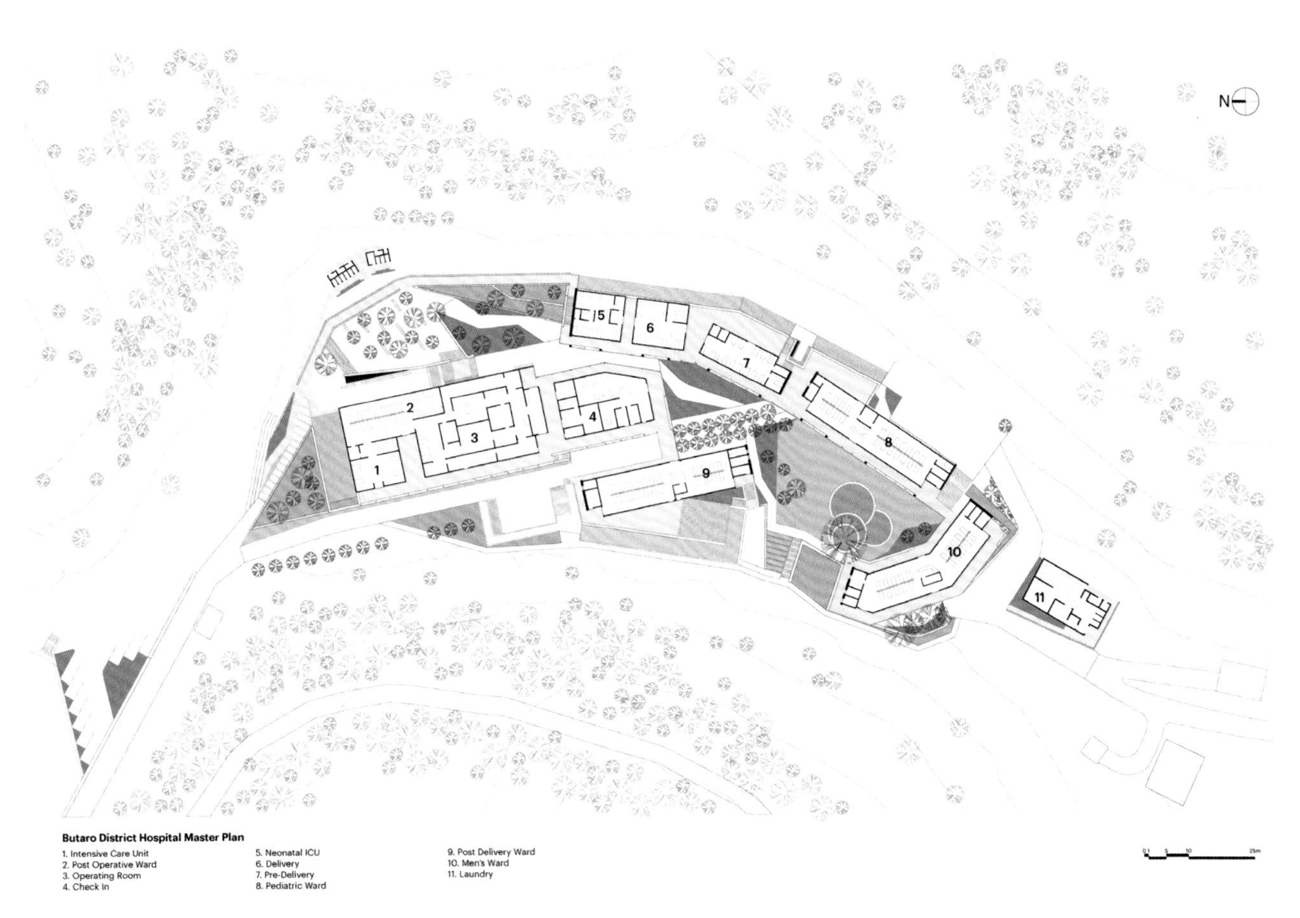

布塔罗地区医院总体规划

1. 重症监护室
2. 术后病房
3. 手术室
4. 登记处
5. 新生儿重症监护病房
6. 产房
7. 待产
8. 儿科病房
9. 产后病房
10. 男病房
11. 洗衣间

▲卢旺达医院总平面图 Butaro District Hospital Master Plan

▼ 卢旺达医院鸟瞰 Butaro District Hospital Aenial View

在布雷拉，人口超过 34 万，非常贫穷。在 2007 年卢旺达卫生和健康组织到来之前，这里没有像样的医院、没有医生。在这里，迈克尔一行与保罗医生团队合力设计了一所 140 个床位的医院。针对之前常见的建筑问题，他们在这次的设计中增加了户外走廊，安装了大半径风扇和紫外线杀菌灯，并使用非渗透性的、连续的地板涂料以保证通风，并减少细菌滋生。

In Burera, the population was more than 340,000, and they were extremely poor. In 2007 before the arrival of Rwanda health organization, this place was one of the last two worst regions in the world where had terrible hospital and no doctors. Over there, Michael and Paul's medical team jointly built a 140-bed hospital, in which decorate with outdoor corridors, large fans, ultraviolet germicide lamps, and floors painted with specific substance maintaining air circulation and reducing bacteria growth.

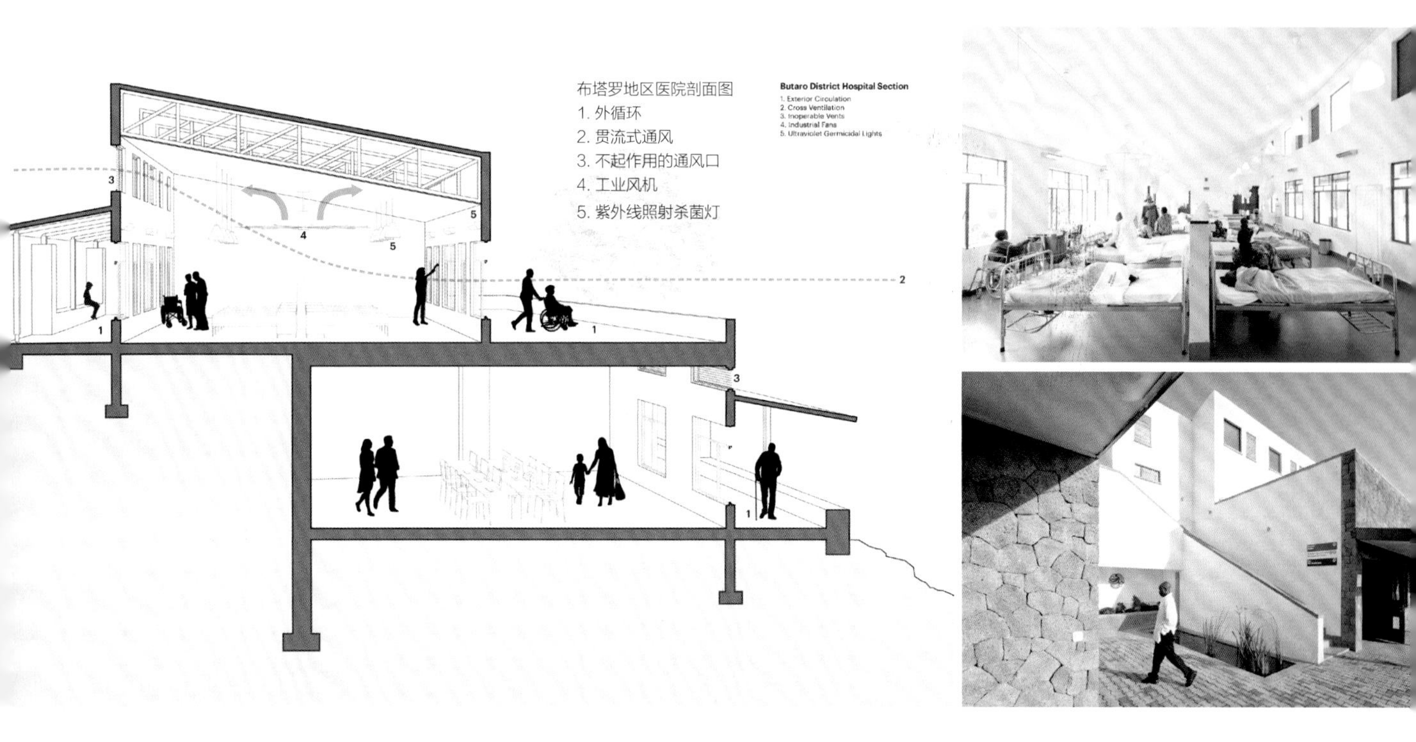

▼ 户外回廊 Outdoor cowidor

▼ 发动民众建设医院 Mobilize people to build hosital

设计方案进入到落地施工环节，如果用常规做法，需要将推土机等重型机械运送到山顶的基地，这意味着巨额的费用。当地工程师布鲁斯·尼泽耶（Bruce Nizeye）提出一种卢旺达理论“Ubedehe”——让“社区”为“社区”服务。几百人带着锄头和铁锹来到现场，用了原计划一半的时间和一半的费用，完成了山顶地基的挖掘和建造。接着布鲁斯又带来了木匠，对群众授之以渔，解决家具购买的问题。在经历了种族屠杀、部分种族灭绝的惨剧之后，共有来自不同性别、不同背景的群众 3898 人次接受了培训，参与了这所医院的建造过程。项目方提供给所有员工食物、水和医疗服务。

The costs of shipping the heavy machineries, such as bulldozers, to the hills should have been huge. Bruce Nizeye, a construction engineer, proposed a Rwanda theory called “Ubedehe”– an idea making “community” service “community.” Hundreds of people came with gimmicks and shovels, and they finished the construction at hills with half time and cost. Bruce then taught to people there to make their own furniture. After the horrible genocide, there were 3898 people with different genders and background participated the entire construction progress, including serving foods, water and medical cares.

在这场行动中，设计者在设计过程中解决了硬性技术问题。而在建造过程中，通过“愚公移山”的方式，让当地的居民在经历了巨大悲痛之后重新团结起来，还节省了建设成本，提供了就业机会，增加了社会效益。迈克尔总结了这一场行动的四大要点：雇佣当地工人、就地取材、培训学徒、给予尊严。

在这所医院建成治愈病人之前，共同建造建筑本身，就已经开始治疗卢旺达居民内心的伤痕了，建造过程令所有人为之振奋与感动。

During this reconstruction, the hardware issues were solved, and the method of “Ubedehe” reunited people there after the tragic grief. It was cost-effective, created job opportunity, and increased social benefit. Michael pointed out the four reasons making this happen, which were “Hire Locally,” “Source Regionally,” “Train Where You Can,” and “Invest Dignity.”

The hospital building, before any patient was cured, was started healing the wounds in Rwandan people, and encouraging them thereafter.

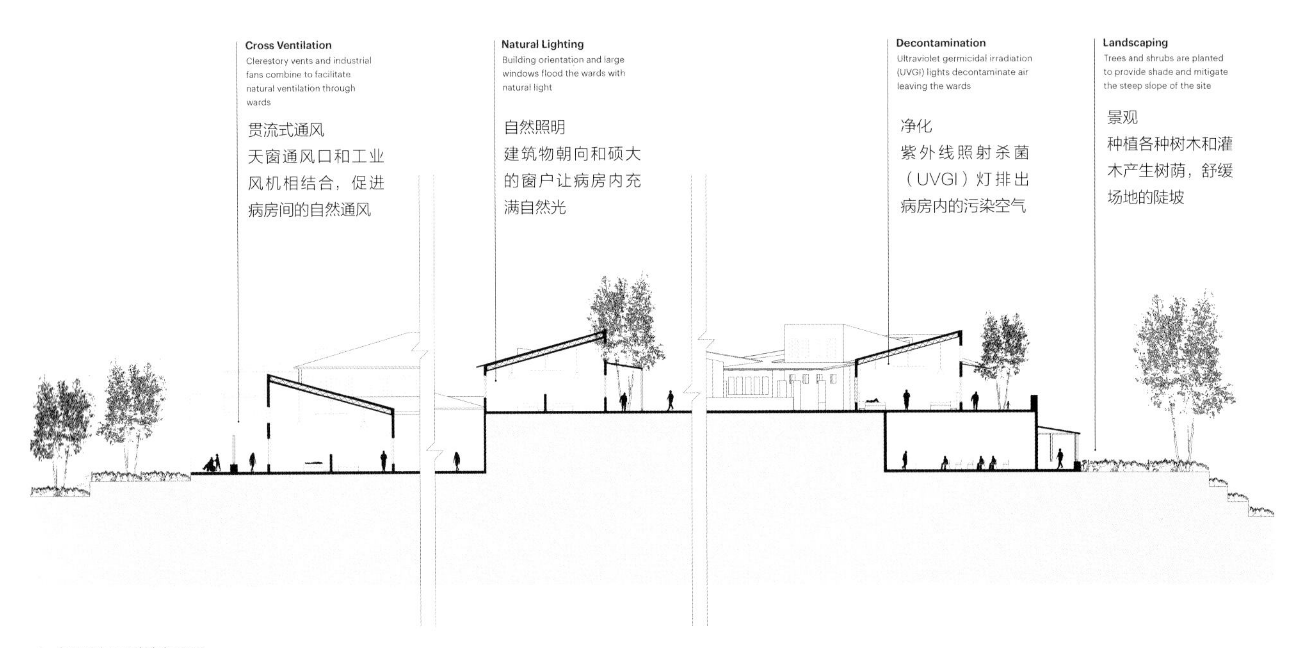

▲卢旺达医院剖面图
Cross–section of the hospital in Rwanda

Mass 建筑事务所的口号之一是“为了社会效益做设计 (Design for Public Interest)”。对于迈克尔而言，建筑设计不仅是通过简单的设计手段达到使用者的功能要求；更重要的是，他能从各种社会活动中看到设计师介入的可能性。在之后的每一次实践中，Mass 建筑事务所都会提出一个设计之外的问题。在海地的 GHESKIO 霍乱治疗中心项目中，他们提问：建筑物是否能帮助抵抗流行病？于是设计了一个让污染性医疗废品在进入水源前被清除的系统，如今已因此救助了许多生命。在马拉维的待产村，他们提问：一个接生中心是否能够从根本上减少母婴的死亡率？马拉维是世界上母婴死亡率最高的地区，Mass 建筑事务所设计了一个可以吸引产妇与同伴提早前来安全待产的接生中心。在刚果，他们提问：是否可以通过一个教育性中心保护濒临灭绝的野生动物？同时偷猎象牙及野兽也是导致全球传染病及战争的原因。于是他们在这个与世隔绝的地方利用当地有限的泥土等材料建设了一个展示如何保护生态物种的多样性的场所……

建筑可以成为变革的引擎。

在迈克尔 · 墨菲及其团队的所有设计工作中，我们可以看到设计思维与人文关怀的实践、符合公众利益的执行探索，以及让当地民众参与的赋权意识。

“Design for Public Interest,” one of MASS slogans. For Murphy, architectural design not only gave what the users’ needs, also discovered the potentials for designers. Every time since that reconstruction plan, Mass Design Group suggested one question other than design. During the project of GHESKIO Cholera Treatment Center, Mass Design Group asked that “whether can the building help defend epidemics,” which was why they designed a cleaning system for medical wastes process before disposing to waster. Many lives were saved therefore. In the village of Malawi, they questioned that “whether can a delivery center essentially reduce the maternal and infant mortality.” Malawi was one of the region in the world with the highest maternal and infant mortality rates. Mass designed a trustworthy delivery center for pregnant women. In Congo, they asked “was it possible to build up an education center that protected the endangered wild species, which were threatened by the illegal hunting that deteriorated the global infection diseases and wars. In the isolated region, therefore, they constructed a building with limited materials that show the way to protect ecological diversity.

“Architecture can be a transformative engine for change.”

We saw the achieved design ideas and humanity cares, out from the design works proposed by Murphy and Mass Design Group, that explore with feasible public welfare and motivated the awareness of the local residences participations.

▼ Mass 建筑事务所为每一个项目提出问题并寻找解决方案　图片来源：Mass 建筑事务所官网
Mass Architecture Firm propose and find solutions for each project.　Photo source：Mass Website

The Butaro District Hospital
Can a building heal?

The Memorial to Peace and Justice
How can a memorial heal our past?

Butaro Doctors' Housing
Can housing help retain doctors?

Butaro Doctor's Sharehousing
Can housing foster a community of care?

Butaro Ambulatory Cancer Center
Can we amplify outcomes of rural cancer care?

GHESKIO Cholera Treatment Center
Can a building help resist an epidemic?

GHESKIO 肺结核医院
设计能否提高肺结核护理水平？

GHESKIO Tuberculosis Hospital
Can design improve tuberculosis care?

Maternity Waiting Village
Can shelter reduce maternal mortality?

The Umubano Primary School
Can design improve access to learning?

2016 年，Mass 建筑事务所携手美国民权主义领袖布莱恩·史蒂文森（Bryan Stevenson）和非营利组织“公正司法倡议（Equal Justice Initiative）”，在阿拉巴马的蒙哥马利设计了“和平与正义纪念馆”（The Memorial to Peace and Justice），为了祭奠过去 4000 多例死于残酷私刑的美籍黑人。建筑的主体悬挂着许多铭刻着逝者名字的柱子，在建筑物外面，若干根柱子等待着，以便刻上那些在炼狱中死去但尚未证明身份的牺牲者。这一次，迈克尔再次拾起卢旺达的建筑模式，联手布莱恩团队收集每一处发生过私刑的当地的泥土，和逝者家庭后代及社区领袖一起把它们封存到各个独立的罐子里。收集泥土的过程本身就是一个心灵治疗的过程——就和当年在卢旺达，民众用双手建造医院，切割火山石的仪式一样。这些泥土将填满每一根柱子，在多年后种出大树，用希望代替悲伤。而最终，这座建筑将会成为一个可以诉说的场所——诉说那些曾经让这片国土伤痕累累的故事——并慢慢地愈合。

建筑不仅仅是雕塑或房子，它让我们个人或集体的愿望浮现于社会，伟大的建筑可以治愈人。

——迈克尔·墨菲

In 2016, Mass and Equal Justice Initiative, found by the civil right leader Bryan Stevenson, worked together in Montgomery, Alabama, where they designed “The Memorial to Peace and Justice” honoring the 4,000 black Americans suffered from brutal lynching. The structural pillars were inscribed with the names of the deceased, and some were even unidentified. This time, following the similar pattern as in Rwanda, Michael and Bryan's team collected the soils on the land wherever the lynches took place. They invited the deceased family and community leaders to seal soils in the separate jars. The process of soil collection was grief-relived that was similar to what happened in Rwanda, where people built hospital with their own hands, and where the lava-stone ceremony was held. The soils would fill every pillar, nurture trees, and replace sorrow with hope. Eventually, the building told the past unspoken stories, where were healing steadily.

“Buildings are not simply expressive sculptures or houses. They make our personal and our collective aspirations visible in society. Great architecture can heal.”

– Michael Murphy

▲ 罐中封存发生每一处私刑的当地的泥土。
The local soil was sealed up in the tank.

▲ 和平与正义纪念馆 2017 年开放。
The Memorial to Peace and Justice opened in 2017.

第2章
CHAPTER 2

SLRD 2017 参展作品
SLRD 2017 EXHIBITION WORKS

SLRD 2017

关于 SLRD
ABOUT SLRD

SLRD 是由四位 80 后青年设计师（卜天静、田甲、孙浩晨、张雷），在邵唯晏先生以及台湾逢甲大学建筑专业学院助理教授兼室内设计学士学位学程主任陈文亮老师的协助下，于 2017 年 11 月发起的一场青年设计师公益行动。

SLRD 寓意为：以设计之力共享爱与尊重（SHARING LOVE & RESPECT with DESIGN）

借由此次“共享”之命题，几位策展人将其作品的主要结构重叠，形成了 SLRD 的 LOGO。这场行动的初衷在于唤起设计人的社会使命感，试图借由设计之力引起关注并尝试解决一些社会现象。他们认为“设计”本身意在呈现美好，而这美好不应只属于少部分有足够经济实力的消费者，那些普通人甚至社会弱势群体往往更需要被关注和优化其生存现状。SLRD 不是昙花一现的口号，他们希望这场行动不仅停留在 2017 年，应有更多有公益热血的设计人加入进来，不断延续，共同探讨。仅以此书记录首届 SLRD 设计行动，并将所有他们在此过程中收获的感谢与感动化作动力，继续前行！

SLRD is a public charity event launched in Nov. 2017 by four post-84s young designers, who were Sissi Bu, Tian Jia, Sun Haochen and Zhang Lei, and was under the assistance from Shao Weiyen, and Chen Wenliang, the assistant professor of the School of Architecture and the director of Interior Design Bachelor Degree Program at Feng Chia University.

SLRD, the abbreviation of Share, Love, Respect and Design, signifies “Sharing Love & Respect with Design.”

By means of presenting the theme “sharing,” these young curators rearranged their design works, constituting the logo “SLRD.” To show their social missions as designers, the curators demonstrated the influence of design in attempt to work out some social issues. They believed that the context of “design” displays the wonderful life, which should not be limited to the affordable consumers, but rather to care and help those minorities. Rather than proposing the slogan “SLRD,” we want to intensify the influence of this event and more enthusiastic designers can join us. We will be continuing and exploring relentlessly. This publishing, as recording our first action of SLRD, transforms all the grateful moment into actions from now on.

策展人介绍 ABOUT CURATOR

卜天静 Sissi Bu

以设计之力影响世界，创造美好，是我作为设计人的心之所向。以设计关注他人，传递内心的能量，令我收获的成就感是任何单纯以商业为目的设计行为都无法比拟的。设计人应当附有使命感及社会责任感。这次行动我们收获了许多的感动，小小的设计也可以温暖人心，鼓舞迷茫中的灵魂，这使我感到无比的荣耀。今后必将沿此道路坚定地走下去，无畏荆棘，如沐春风。

Influencing the world and creating the wonderful life with "design" is my goal as a designer. Caring people with "design" and transmitting the inner power make me feel more fulfilling than does any other business-oriented design. Designers should have missions and responsibilities to society, and any little design can be inspiring and encouraging the lost. That was such a great honor, and I will pursue it firmly and ceaselessly. The journey of design is fearless and joyful.

田甲 Jia Tian

"设计"之于我，是空间与人的感受，我相信用空间的力量可以改变世界，设计可以令平凡未知，当今世界每一件小事都有重新思考一遍的必要性。因此呼吁每个设计师不要仅仅为了利益而设计，更要为了梦想而设计。新一代的设计师不是残酷生活的行尸走肉，每个人生而不凡，应竭尽全力致力于世界美好，我们坚信星星之火可以燎原。

For me, "design" is the feeling of space and people. I believe that the power of space can change the world. Design creates the opportunity to give a second thought to any tiny, ordinary or unknown things. Therefore, designers should design not only for the profits, but for the dreams. A new generation of designers are not the walking dead for the cruel life. Every life is one-of-a-kind, which should be committed to making the world better. We firmly believe that a single spark can start a prairie fire.

孙浩晨、张雷 Haochen Sun, Lei Zhang

其实我们更关注人和空间的关系—— 一种氛围，而非空间本身的实体感。人和人的行为是设计中的一个很重要的因素，所以我们力求体现人们在空间中有怎样的体验，以最直接的方式来表达想法而不是通过具体的形象或者形状。方案的推敲过程中我们尽量选择有最多可能性的方向。设计的可能性对我们来说也意味着多样性和灵活性的增加。

In fact, we pay more attention to the relationship between people and space, which is the ambience rather than the physical sense of space. Human behavior is an important factor in our design. So we have been trying to present people how they feel in the space, presenting by means of the candid methods, rather than solid images or shapes. We try to choose the direction with the most possibility in the process of the scheme. The possibility of design also means an increase in diversity and flexibility.

DESIGNER

中国
卜天静 Sissi Bu

中国
田甲 Jia Tian

中国
张雷 Lei Zhang
孙浩晨 Haochen Sun

中国
王昊 Wang Hao

马来西亚
张绩 Joshua Teo
中国
张皓翔 Jack Chang

中国
闵滢如 Lvy Min
陈又宁 Annie Chen

中国
黄华敏 Min Huang
高语辰 Eunice Kao

中国
曾子麟 Andy Tsang

中国
留鸿运 Elvis Liu

中国
施嘉晓 Sison Shi

马来西亚
邓元斌 Jesden Tang

马来西亚
叶铭盛 Sean Yap

勇气
NERVE

卜天静 Sissi Bu（中国）

一个作品并不能从根本上改变一个时代的进程，也不能消除世界上的悲剧。但至少，设计师希望通过这个作品，可以让更多的人通过直观的体验，关注到这个时代坚硬外壳下面隐藏着的另一面。在这个重压的时代，精神健康已成为严重而又耗资巨大的全球性卫生问题，我们无法继续视而不见！

Nerve n. 神经；勇气

vt. 鼓起勇气

你有足够的勇气吗?

你有胆量来碰触现实吗?

你有那一根痛苦的神经吗?

A design project cannot change the track of an era. Nor can it eliminate tragedies in the world. But at the very least, the designer hopes that, through the intuitive experience of NERVE, more people will pay more attention to the truth hidden under the hardshell of our era.

Nerve n. (1) Nerves are long thin fibres that transmit messages between your brain and other parts of your body. (2) Courage

Vt. To encourage sb to be confident

Do you have the nerve?

Do you have the nerve to face the reality?

Do you have painful nerves?

▲ 作品“勇气”实景图
Reality images of NERVE

设计背景 | BACKGROUND

曾经，心脏病、艾滋病、霍乱……这些疾病令人闻之色变，一旦患上，几乎等于死亡。而今天，这些疾病的死亡率已经减少了70% 以上，一个 20 岁左右的艾滋病病毒携带者，很可能要到他六七十岁甚至更晚的时候死于其他老年并发症。这些疾病患者寿命的延长不仅得益于医疗技术的进步，同时也得益于社会认知的普及。而心理、精神疾病的治疗技术及大众的社会认知没有那么快的进步，所以许多精神疾病患者还在忍受方方面面的不理解，有些病症甚至很晚才被发现。

Once upon a time, the attacks such as asheart disease, AIDS and cholera were intimidating, almost synonymous with death. But today, the death rates of these diseases have been reduced by over 70%. A patient in his/her twenties carrying AIDS virus is still very likely to live till his/her sixties, seventies or even longer. The success in curing these diseases is attributable not only to the medical technologies progress but also to the popularization of social recognition, high-priority attention, and early detection. However, progress in psychological/mental illness treatment has been relatively slow because of various misunderstandings, ignorance, neglect, extremely late detection, and even social discrimination and mockery.

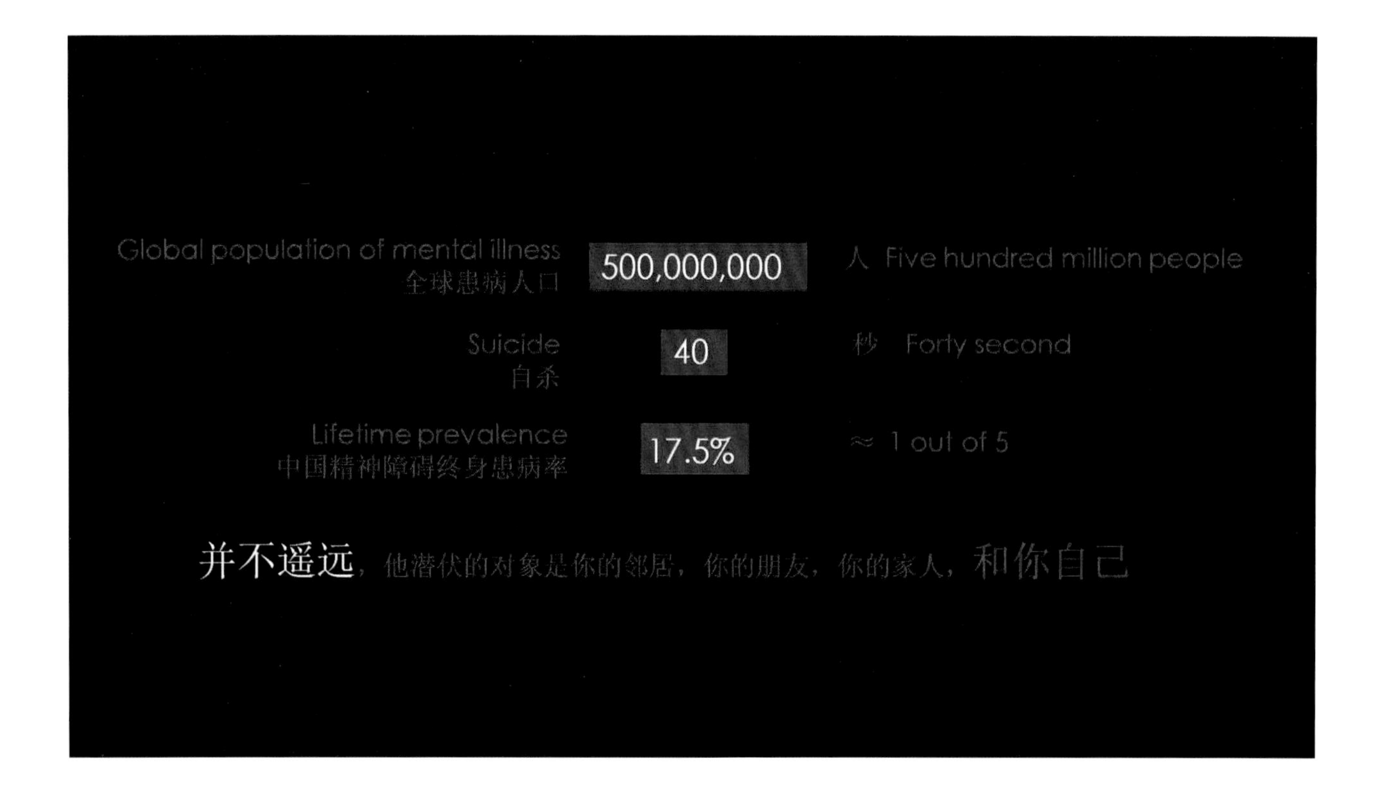

根据世界卫生组织的统计，全世界有将近五亿人受到精神疾病的影响，每 40 秒就有一人死于自杀。精神健康障碍已经成为严重而又耗资巨大的全球性卫生问题，影响着不同年龄、不同文化、不同社会经济地位的人群，在中国各种精神障碍的终身患病率是 17.5%。

The World Health Organization estimates that there are nearly 500 million people suffering from mental illness, and a suicide is committed every 40 seconds. Mental health disorder has become a serious and extremely costly global health problem, affecting people of different ages, cultures, and socioeconomic groups. In China, the lifetime prevalence rate of various mental disorders is 17.5%.

约翰 · 海杜克（John Hejduk, 1929—2000），建筑理论家和历史学家塔夫里称他为“纽约五人组”中最重要的人物，也被誉为当代美国建筑的“四教父”之一。

1967 年他开始探索研究，利用几何形状来进行空间创造的练习，用对角线和曲线网格打造更严格的细节，但是他很快就转移到了一个更加“自由”的方法。他开始探索建筑设计的新影响:心理学、神话和宗教。他画的图纸往往会通过一个黑暗的镜头展示建筑的主题，他最有名的新英格兰“假面舞会（1981 年）”就是受到异化婚姻的启发。

约翰 · 海杜克的作品往往具有戏剧性的张力。他给“建筑”注入性格，塑造为“角色”与城市、体验者对话，并透过建筑探索更深层次的哲学问题。受其启发，本案设计师希望创造一个“意识空间”，让空间体验者关注到一些常被忽略的社会现象，并且思考和自省。

John Hejduk (1929 – 2000) – Architectural theorist and historian Mafredo Tafuri said that John Hejduk was the most important figure of the “New York Five”, and John Hejduk was praised as one of the “four founding fathers” of contemporary American architecture.

In 1967, he began his early exploratory research which involved the creation of space using geometric shapes, but he soon moved away to explore new influences in architectural design: psychology, mythology, and religion. His drawings often expressed themes of architecture through a rather dark lens, and his most famous work, the New England Masque (1981), was inspired by dissimilated marriage.

John Hejduk’s works often possess dramatic strength. He characterized architecture as vivid roles that can converse with the city and people, and tried to explore deeper philosophical questions through architecture. Inspired by John Hejduk, designer of this project wanted to create a “consciousness space” that can remind people of some serious social issues that are often neglected and lead them to think and introspect.

▲ 约翰 · 海杜克 John Hejduk

▲ 约翰 · 海杜克的设计图 John Hejduk’s design

▲ 约翰 · 海杜克手绘作品 John Hejduk's hand-painted works

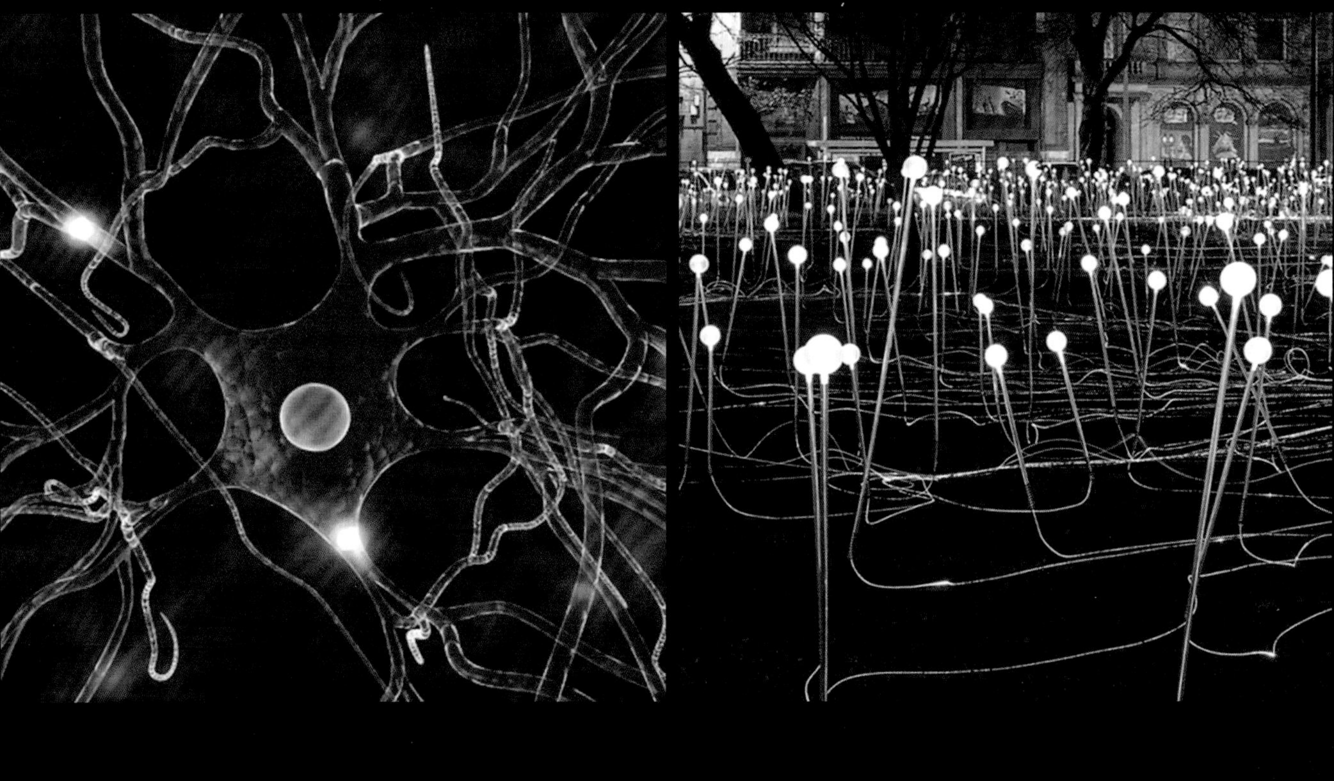

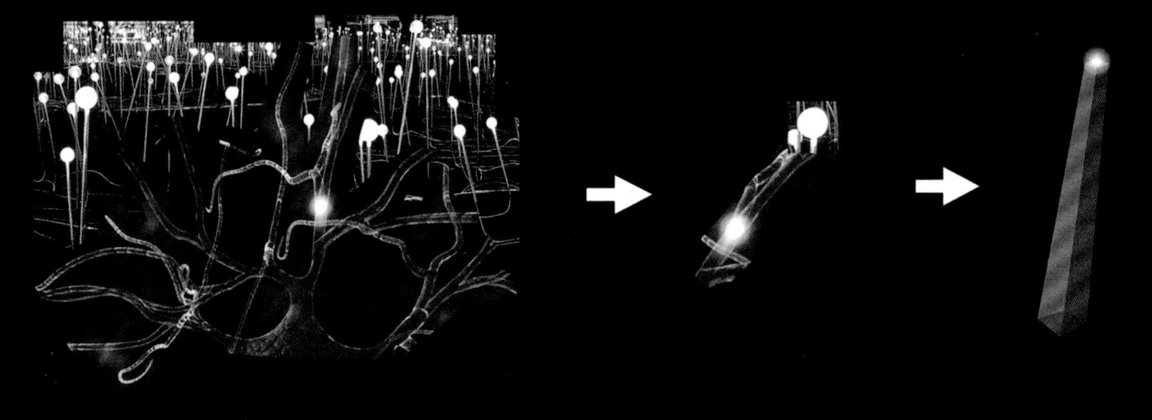

NERVE CELL + FIELD OF LIGHT

神经细胞 + 光场

COMBINE UNIT

组合单元

NERVE UNIT

神经元

设计说明 | DESIGN DESCRIPTION

▲ “勇气”效果图　Design sketch of NERVE

作品结构体外立面的主要材质为做旧钢板，与内部柔软的海绵材质形成对比。900 根不同长度及方向的海绵是借由 GH 数字语言写出的结果。当触点靠近时，这些触角会随之改变形态，仿佛有生命的神经被靠近和触动。

外壳黑漆漆的钢板，密不透风，犹如监狱和牢笼。也像一个面具，人可以躲进里面，伪装自己，拒绝袒露内心柔软的部分，也害怕接受来自外界的阳光。侧壁小小的方洞，是唯一可以连接内外的窗口。

The façade of NERVE is mainly built by steel plates, which are in strong contrast to the soft sponge material of inner space. 900 “Nerve antennae” of different lengths and directions, were created by digital language programed through “Grasshopper”. In the GH program, these “tentacles” change their lengths and shapes when a virtual cursor draws near, , just like live nerves that change when something is approaching them.

The “shell” of NERVE is made of black steel plates that are airtight as if it's a prison or a cage. In the designer's eyes, it's just like the mask that people wear in modern days in order to disguise themselves, hide their internal tenderness, and avoid sunshine from the outside world. The small square hole in the wall is the only window that you can peek into the inner and outer space.

通过计算机的参数式运算及互动装置的搭配，创造出具有互动性的心灵空间。

Using computer's parametric computing and interative installaction to create an interactive installation for spiritual space.

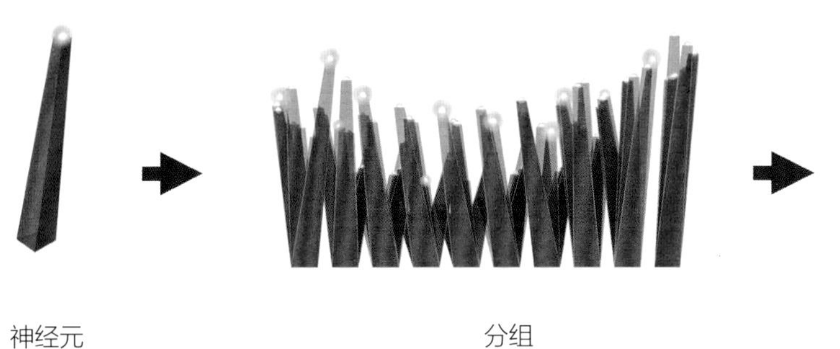

神经元

NERVE UNIT

分组

GROUPING

调整长度和角度

ADJUST LENGTH & ANGLE

组合生物电传感器

COMBINE BIOELECTRIC SENSORS

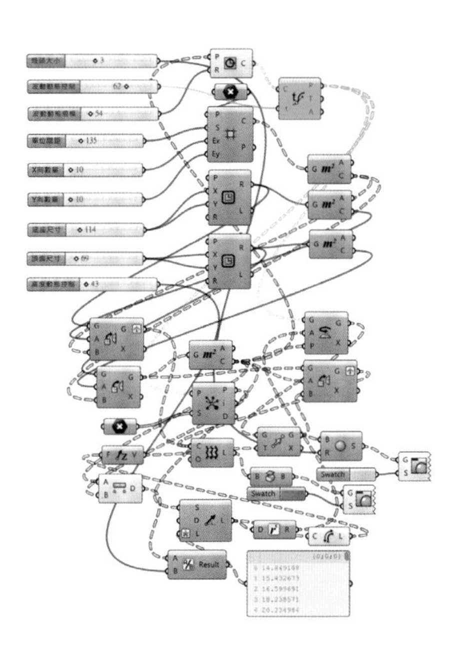

拉开两块厚重的布帘，犹如舞台拉开了帷幕，内部生长着将近 900 根不同长度的红刺，每一根红刺的末端都镶嵌着星星点点的光纤灯。狭窄而曲折的走道被红色芒刺包围着，暗示着一段略带危险的旅程。天花上的镜面将空间无限延伸。身置其中，体验者也许会感到疑惑，这个空间到底在诉说着怎样的故事?

体验者在狭窄的走道中穿梭， 其中 16 根最长的触角末端安装了触摸条传感器，碰触到它们的时候，敏感的“神经”被触发，陆续响起各种不同的声音——警笛声、脚步声、急促的砸门声、喘息声……这些看似平常的声音，描绘着一个精神障碍患者日常独自在家时进入极度紧张和恐惧中的状态。

心绪未定，体验者钻出黑匣子，顺梯子向上爬， 二楼的洞口处，一个镜面展演室播放着本装置设计的初衷。触目惊心的片段，是大部分患有心理、精神障碍人群的现实：内心遭受着病痛，却常常难以得到理解、包容，没有倾诉的出口，在这个社会中被疏离、嘲笑和指责。

The stage kicked off when you pass through the massive curtain. There are nearly 900 “Nerve antennae” made of sponge of different lengths and angles inside NERVE, antenna has in-laid sparkling optical fiber lights at the end. These red thorns surround a narrow, meandering path, imply a slightly dangerous journey. The ceiling mirror gives unlimited extension to the space. Standing inside this space, one may be puzzled: what kind of story is this space trying to tell?

Shuttling through this narrow path, you can find 16 Touch Bar sensors installed on the ends of the 16 longest “Nerve antennae ”. The sensitive “NERVE” will be triggered when you touch them, and different sounds are released: alarm, footsteps, door knocking, wheezing, etc. These seemingly ordinary sounds could cause a mental-disorder patient feel extreme tension and fear at home alone.

Walking out of the “black box” with undetermined mood, when you climb up the ladder on the back of NERVE, you will find a mirror media room on the second floor. The video shows the truth of this installation. Those terrifying scenes are the reality that most psychological/mental disorder patients have to grapple with. They suffer not only from illness but also from people’s misunderstanding, intolerance, and refusal to listen. They are marginalized, mocked, and condemned in society.

▲ 勇气二层放映的短片《现实还是幻觉》镜头
Storyboard of short film Reality or illusion played in NERVE

▲ 勇气二层放映的短片《现实还是幻觉》镜头

Storyboard of short film Reality or illusion played in NERVE

▼ 勇气二层放映的短片《现实还是幻觉》镜头
Storyboard of short film Reality or illusion played in NERVE

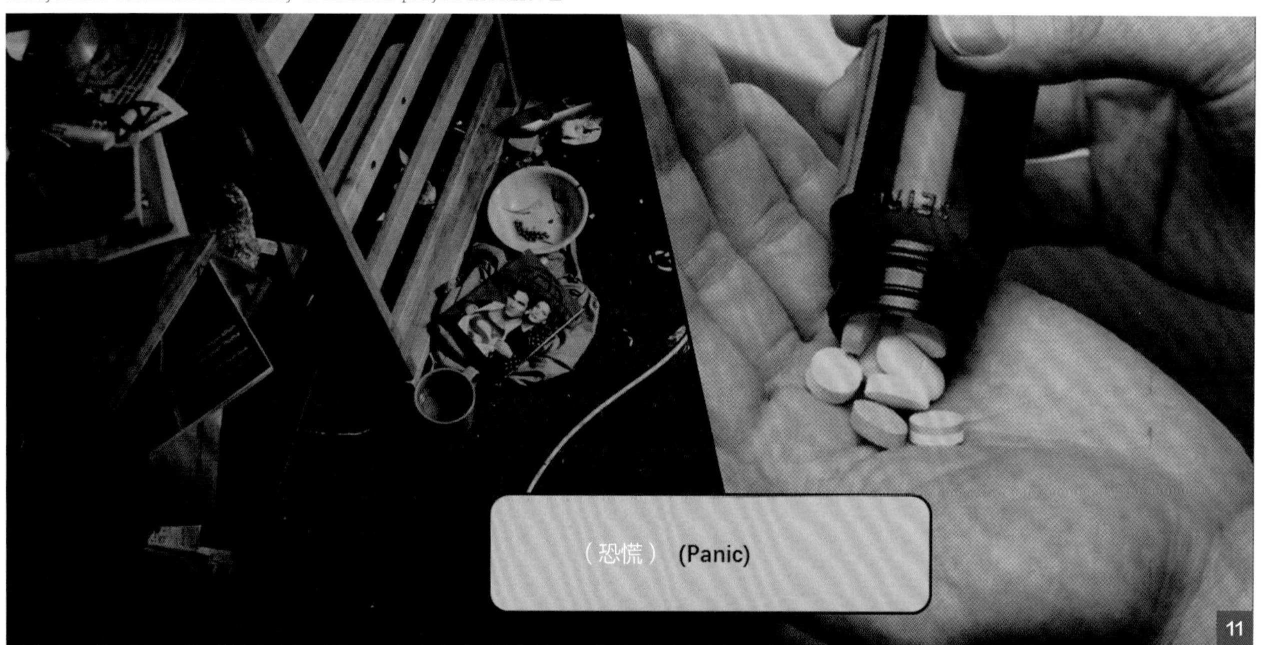

视频部分原始素材来源于纪实摄影师 Jim Mortram。在他独自照顾患病的母亲长达 15 年之后，他感到自己似乎完全与外面的世界隔离了，充满了畏惧和忧郁。

长达一年的时间，他甚至都没有大声说过话，饱受焦虑和抑郁之苦。有一天，一位校友借给他一部相机，摄影让他看到了生活的希望，学会了解过去，立足现在和憧憬未来。

The original material of the video partly comes from the documentary photographer Jim Mortram. After taking care of his sick mother alone for 15 years, he felt as if he was completely isolated from the outside world, often filled with fear and sadness.

For more than a year, he did not even speak aloud, suffering from anxiety and depression, one day an alumnus lent him a camera, Photography enabled him to see the hope of life, to learn to understand the past, to be based on the present and to look forward to the future.

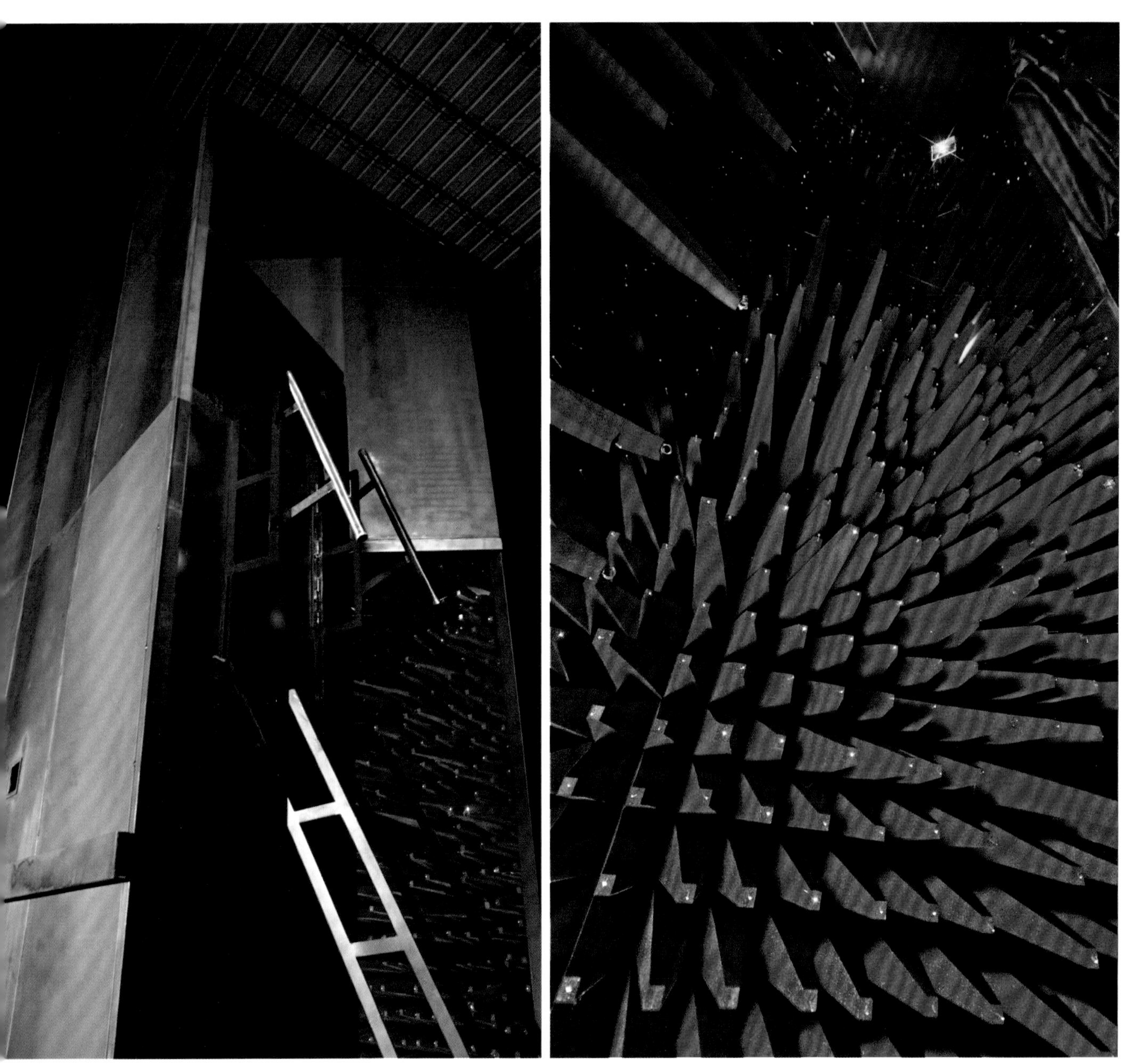

▲ 作品《勇气》实景图　Design sketch of NERVE

▲ 作品《勇气》实景图　Design sketch of NERVE

“我们能做点什么？”

“如果有一天，我也变成这样怎么办，会不会也被抛弃？”

本设计不能给出答案，但每个人都会有自己的答案。

NERVE 原本的意思是“神经”，但设计师更喜欢它的另外一个注解——勇气。

“What can we do?”

“What if I become one of them someday? Will I be abandoned just like them?”

“NERVE” cannot give you the answer, but everyone has his/her own answer.

“NERVE” originally means the nervous system, but the designers prefer another definition of this word – courage.

▲ 作品《勇气》实景图　Design sketch of NERVE

建造过程 | CONSTRUCTING PROCESS

▲ 建造过程　Construction process

一个作品并不能从根本上改变一个时代的进程，也不能消除世界上的悲剧。但至少，设计师希望透过“勇气”，可以让更多的人通过直观的体验，关注到许多人在坚硬外壳下面隐藏着的另一面。我们不能袖手旁观，他们需要正确的认知，请大家一起携手，关注并关爱心理、精神障碍人群。

A design work does not fundamentally change the progress of an era, it also cannot eliminate all the tragedies in the world. At least, by NERVE, I hope that more people will be able to pay more attention to the issues being swept under the rug that is mental health issues.We cannot continuously be a bystander on such pressing issues. On the contrary, we must join to care for people with mental disorders.

▲ 建造过程　Construction process

▲ 精神科专家 Vikram Petal 医生，联合了几个不同国家的医生做了一个实验。在社区中为普通人群做专业的心理指导，并在当地对患有产后抑郁等心理疾病的患者进行干预治疗。实验数据表明，这些以医疗志愿者辅助患者的形式，治愈率比专科医院高出近一倍。设计师希望通过设计作品“勇气”，在社区中产生口口相传的效果，让更多的人正视和关注心理及精神疾病

Psychiatrist Dr. Vikram Petal, teamed up with several doctors from different countries to do an experiment – provide professional medical guidance to normal people in community, give health interventions to patients with psychological diseases such as postpartum depression. Experimental data showed that the cure rate of these non-professional volunteers is nearly double that of specialized hospitals. The designers hope to spread the consciousness in the community, so that people can learn more about mental illness, rather than just few medical facilities

▲ 作品《勇气》实景图　Design sketch of NERVE

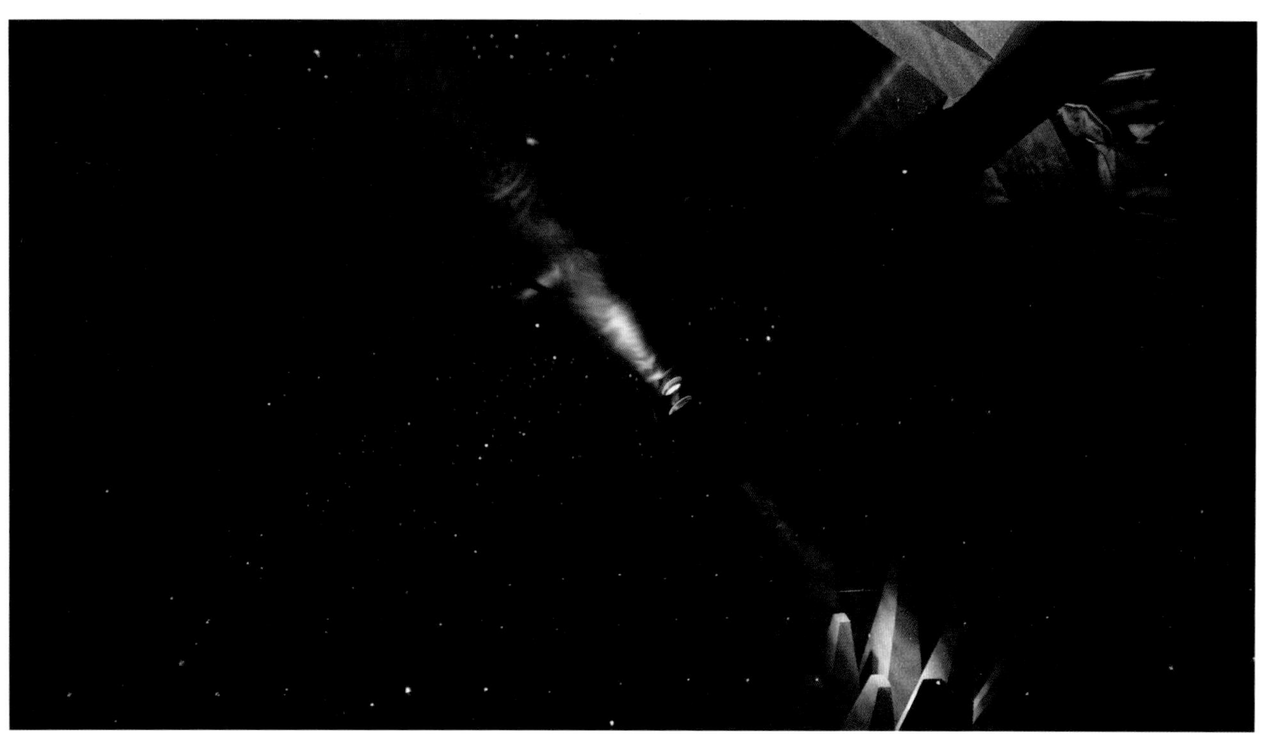

▲ 作品《勇气》实景图 Design sketch of NERVE

社会微救助

SOCIAL MICRO ASSISTANCE

田甲 Jia Tian（中国）

当今世界的扁平化，已经是一个不可逆转的趋势。然而在世界扁平化进程中身为设计师的我们，又能为这个世界做些什么？这一直是我思考的问题。社会被救助人群主要有流浪人群、拾荒人群、乞讨人群三大类，他们成为当今城市中不得不面对的社会弱势群体。有的人对他们心生怜悯，希望给予他们帮助，有的人对他们漠不关心。事实上，社会被救助人员如果得不到有效救助，对社会来讲也是危害社会安定的不良因素。

此次共享空间设计，是从一张最简单的二维木板出发，打造一个三维的微型社会救助空间。这个项目的概念和构想，一是考虑便于运输和搭建，二是考虑材料本身的成本，要便于社会推广。一张板通过折叠打开，像一张折叠椅一样，就满足了对于微救助空间的家具、功能、空间的需求。这样的一个二乘以二的微救助空间不仅是一个装置设计，也是一个可以落地的共享空间。

社会微救助空间，还将继续推出功能更加完善的 2.0 版本。主要会从回归生活、回归社会、回归家庭三个方面来实现被救助人群的社会回归属性。从逐步实现自给自足到被社会认可，再到自食其力地重建家庭，是社会微救助的最终目的和目标。

The world is flat, which is irreversible. I have been thinking about what can we do as a designer in the progress of flattening world. The social subsidy is mainly granted to those homeless, scavengers and beggars, who have become socially vulnerable and the victims under the progress of city development. Some people show their compassion for these helpless, nevertheless, some show their indifference. In fact, if the helpless are not effectively helped, they might be threatening to the social security.

The sharing space design creates a three-dimensional space of social assistance by using a two-dimensional board. The concepts and ideas, firstly, are easily built up and transported. Secondly, the cost reduction of materials enhances the public accessibility. The foldable board, figuratively as a folding chair, is the example of social micro assistance in meeting the demands for furniture, function and space. Such a 2 x 2 micro assistant space presents to people not only a installation design, but also a practical sharing space.

The promotion of social micro assistant space will launch a more perfect version 2.0, which focuses on helping these socially vulnerable come back to life, to society and to home that bring back on track of the social order. The ultimate goal of social micro assistance is to accomplish self-sufficiency, which sequentially leads to social recognition and healthy family life.

一张木板通过折叠旋转，完成形式的转变，从而也满足了对于空间功能的需求。

A foldable board has successfully transformed into a different format, and has successfully met the needs for space functionality as well.

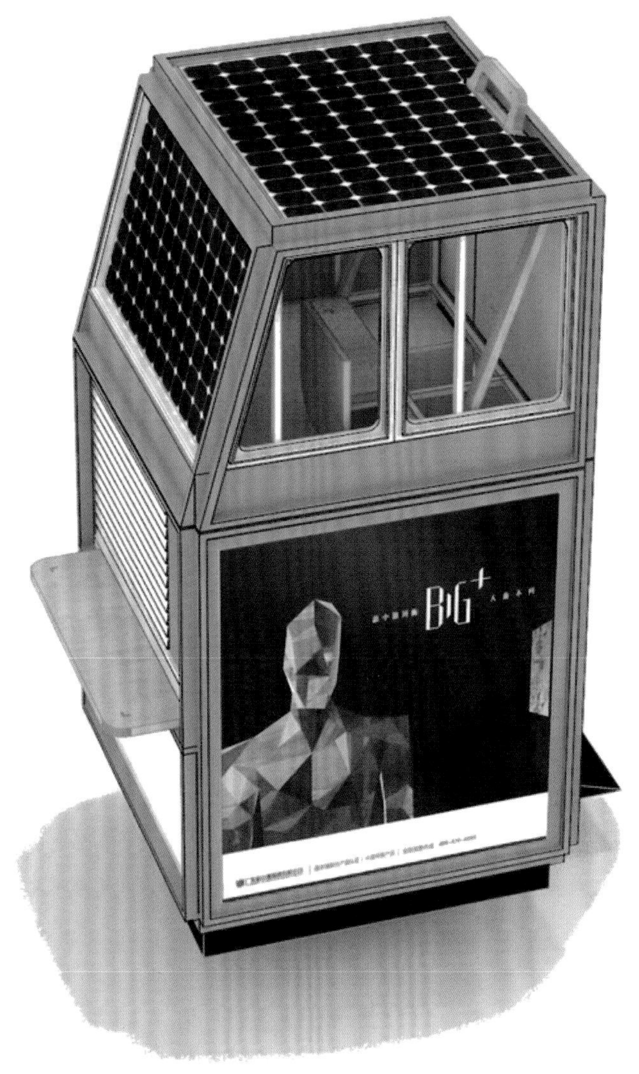

▲《社会微救助》效果图　Design sketch of SOCIAL MICRO ASSISTANCE

▲《社会微救助》的目标群体 Target group of SOCIAL MICRO ASSISTANCE

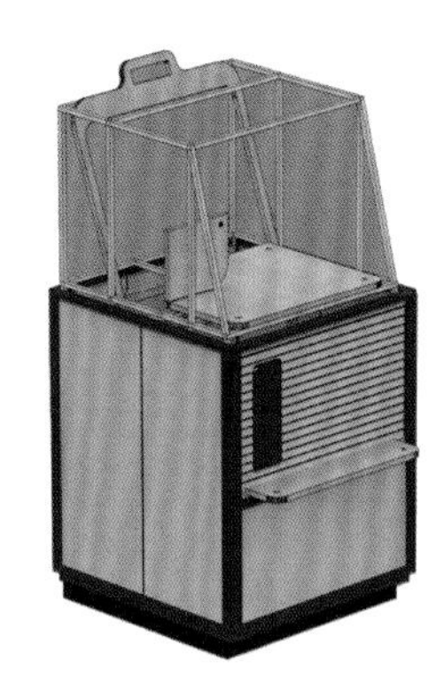

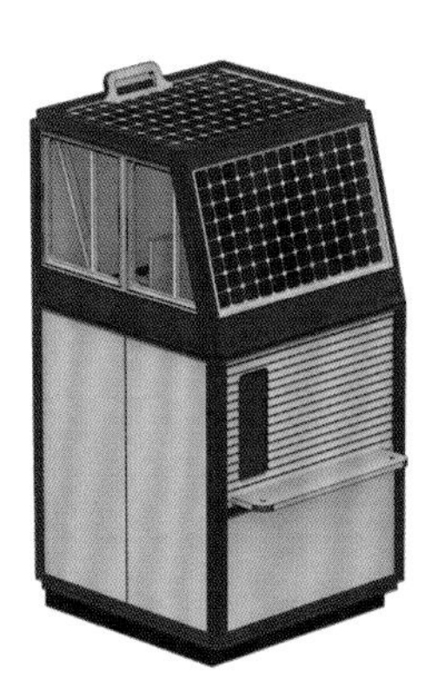

成本再降低　二次利用　更好地推动和复制社会微救助
COST REDUCTION, REUSE, DURABLE

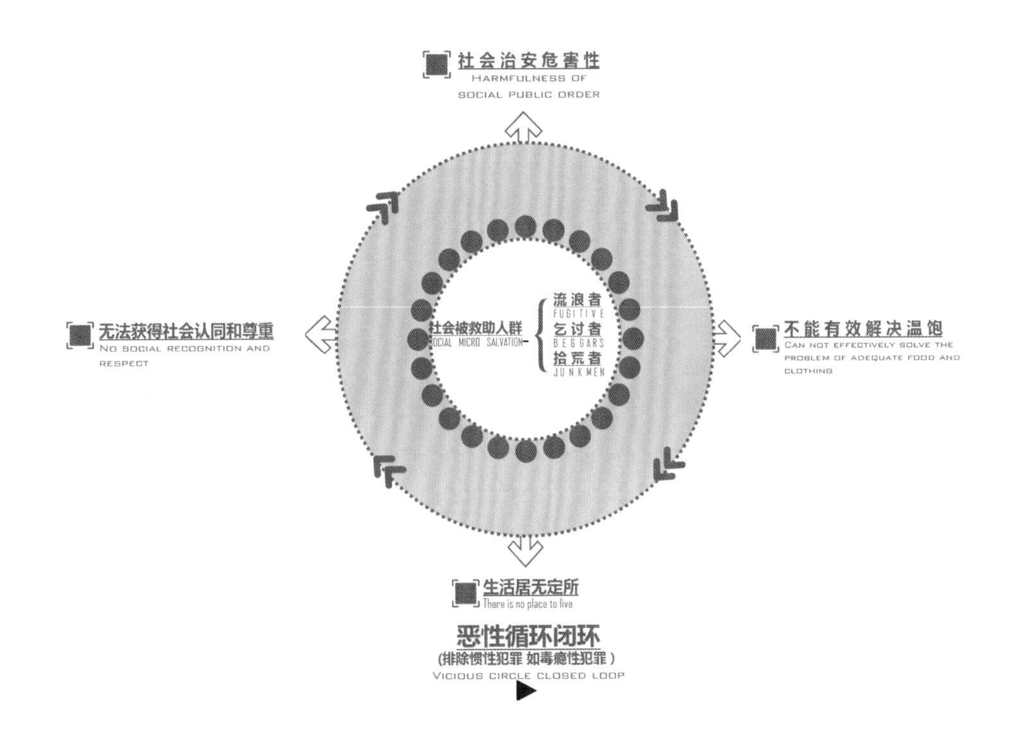

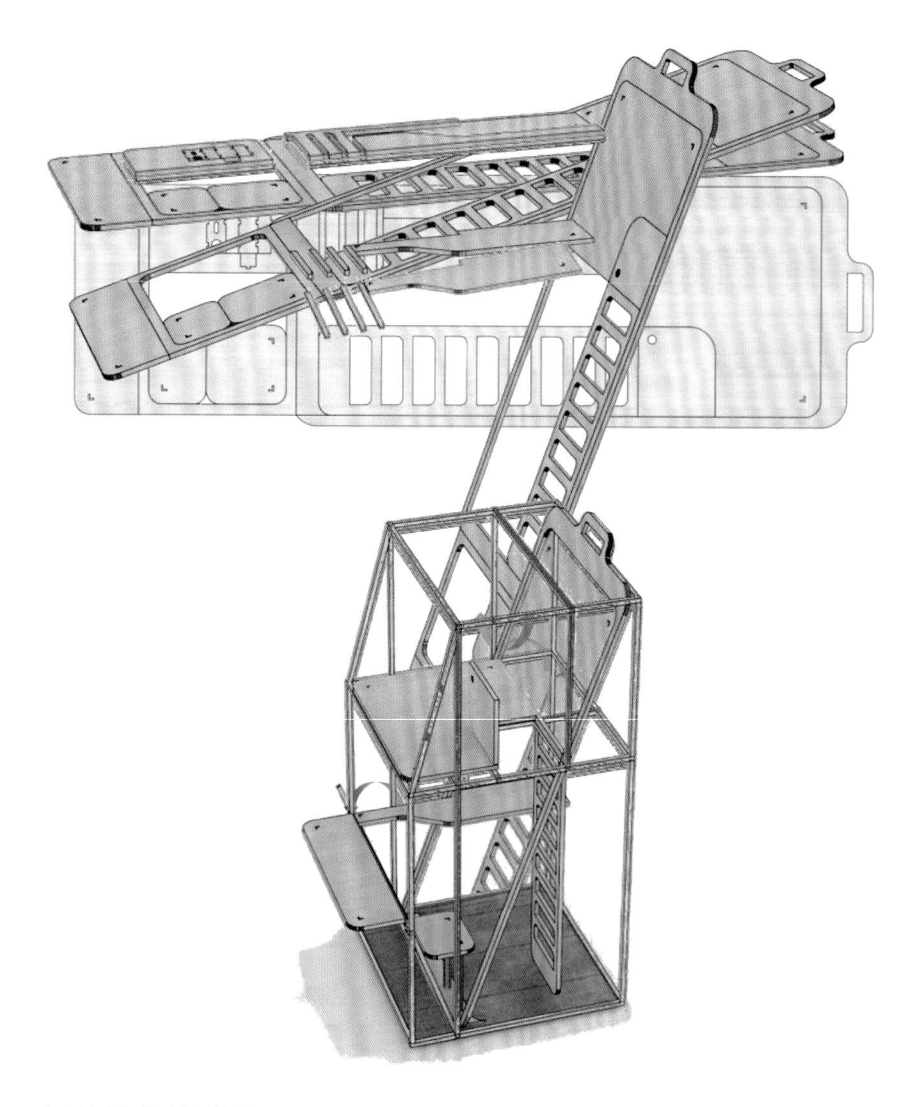

▲《社会微救助》构造图

Structural diagram of SOCIAL MICRO ASSISTANCE

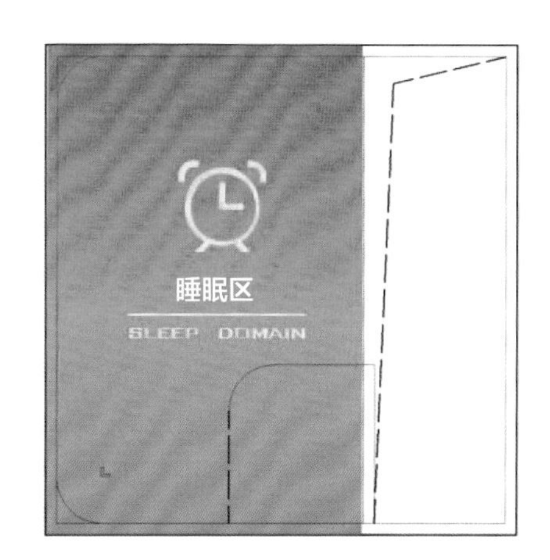

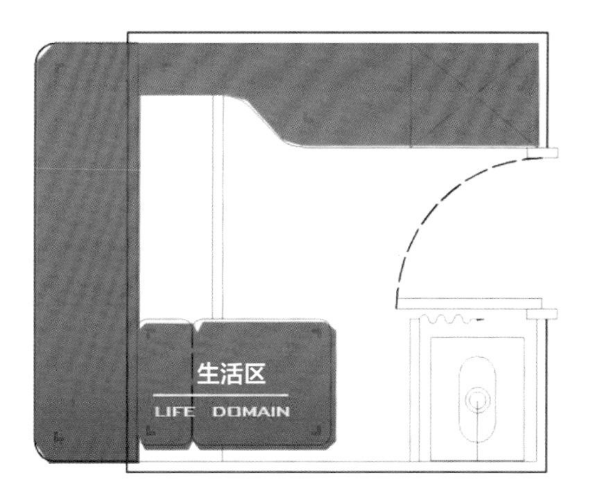

▲《社会微救助》示意图
Schematic diagram of SOCIAL MICRO ASSISTANCE

▲ 建造中的《社会微救助》
SOCIAL MICRO ASSISTANCE under construction

▲ 概念图分析
Analysis of conceptual graph

知识共享
UNIT X

孙浩晨 Haochen Sun（中国），张雷 Lei Zhang（中国）

知识和阅读空间一直是我们研究的议题。Unit X 是一个可存在于不同空间的知识分享枢纽装置。设计师想借此表达的核心诉求是分享知识、创造阅读及相关活动。它由层叠交错的木板组成，阶梯式书架呈现出要邀请人们走上书架阅读书籍的姿态。其中的单元盒可以自由组合成桌、椅和所需活动的形式，不同的组合将创造独特的空间组合方式。设计师经过大量调研，观察人们在公共社区的活动和行为，尝试创建一个让人们能产生互动的设计，在一天的不同时间支持许多不同的知识创造及分享活动。Unit X 拥有数百本书籍和不同的互动装置，不定期的更换为 Unit X 带来新的活动，如读书、演讲、交流会、插花、绘画、小型音乐会、电影放映、民族手工艺展示等。”同时，Unit X 以其低廉的成本及快捷组装方式适用于公园、学校、商场、中庭、广场等不同场地。而其中的互动影音装置将各种知识生产者的活动通过互联网分享给全世界的同时，Unit 协会也将与支教 2.0 联盟，给教育贫乏地区的孩子们带来沟通无极限的理念，开阔孩子们的视野，让他们与世界各地的活动互联，让他们在学习知识的同时也能感受到更广阔丰富的世界。

Knowledge and reading space have always been our research topics. With the appearance of a bookshelf, Unit X is a knowledge sharing hub that can be placed in different functional spaces. The essential idea of the project, according to the designers, is to create spatial experiences conducive to reading and knowledge-sharing. Made of laminated, interlaced wooden boards, the stair-shaped bookshelves are ready to invite people to walk on and read. The unit boxes can be freely assembled into tables and chairs as required by different activities. Hence variations in unit combination perfectly responds to variations in spatial configuration. After a long period of research, in which the designers observe the activities and behaviors of people in communities, the project aims to create a installation that allows people to interact, and support many different knowledge-creating and sharing activities at different times of the day. "Unit X contains hundreds of books and different interactive devices. The occasional replacements of the books and devices will brings new action possibilities to Unit X, such as reading, speaking, exchanging meetings, flower arranging, painting, small concerts, film screenings, ethnic handicraft exhibitions, etc." says one of the designer. At the same time, with its low cost and quick-assemble capacity, Unit X is ideal for medium and large public spaces such as parks, schools, shopping malls, atrium, plazas, etc. The various activities of all these ‘knowledge producers’ will be shared throughout the world by the interactive AV recording devices. The sharing will also provide children in less-developed area with innumerable ideas, which will not only broaden their experience, make them more connected to the world, but also encourage them to interact with people from all over the world and better engage in practical activities.

设计团队选择的木材都是环保桦木多层板，在满足长期使用的同时也能保证其柔和质朴的氛围。整个书架用 106 块垂直木板和 55 块水平木板搭建而成，46 个盒子可以自由移动和组合，所有的材料都可回收利用。

The design team selects environment-friendly birch multilayer board for the constructing material. It not only satisfies the long-term use, but also serves to create a soft and simple atmosphere. The whole shelf was built with 106 vertical boards and 55 horizontal boards, and the 46 BOX can be moved and re-assembled effortlessly, and all the materials can be recycled.

▲《知识共享》实景图
Reality images of UNIT X

这个装置将成为探讨当今知识共享的平台，承担起临时性公共知识分享枢纽的功能。

This device will serve as a temporary knowledge sharing hub as well as provide a platform on which topics about contemporary knowledge sharing will be dicussed.

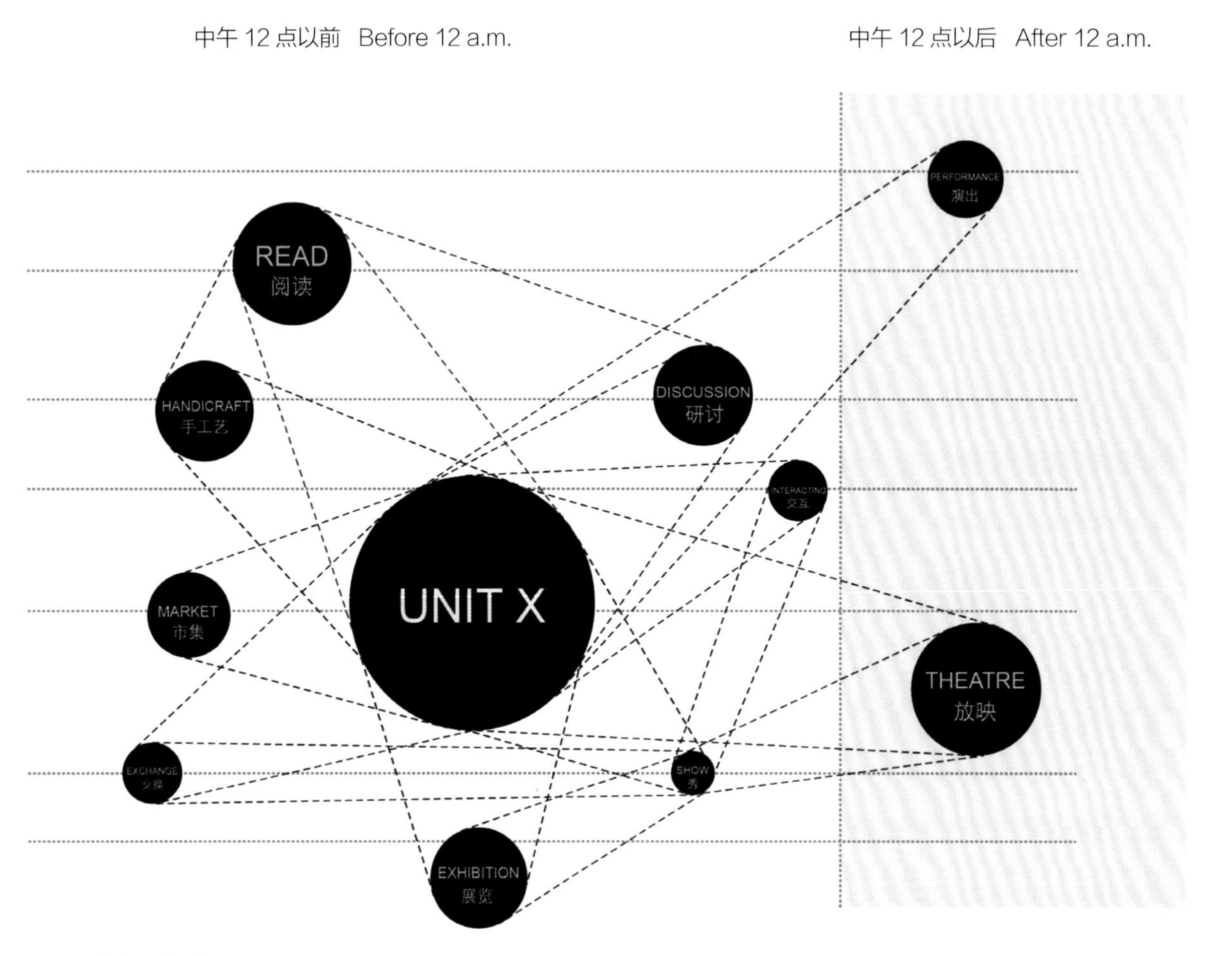

▲《知识共享》功能介绍
Function Introduction of UNIT X

在孙浩晨和张雷看来，知识共享空间的存在，是超越固定且单纯的生存方式的必要选择，为人们提供有意义的生活条件的一种方式。

According to Unitx designers Haochen Sun and Lei Zhang, the existence of knowledge-sharing spaces gives us a way to go beyond the fixed and simple living style that merely focus on necessities. It provides people with meaningful living conditions.

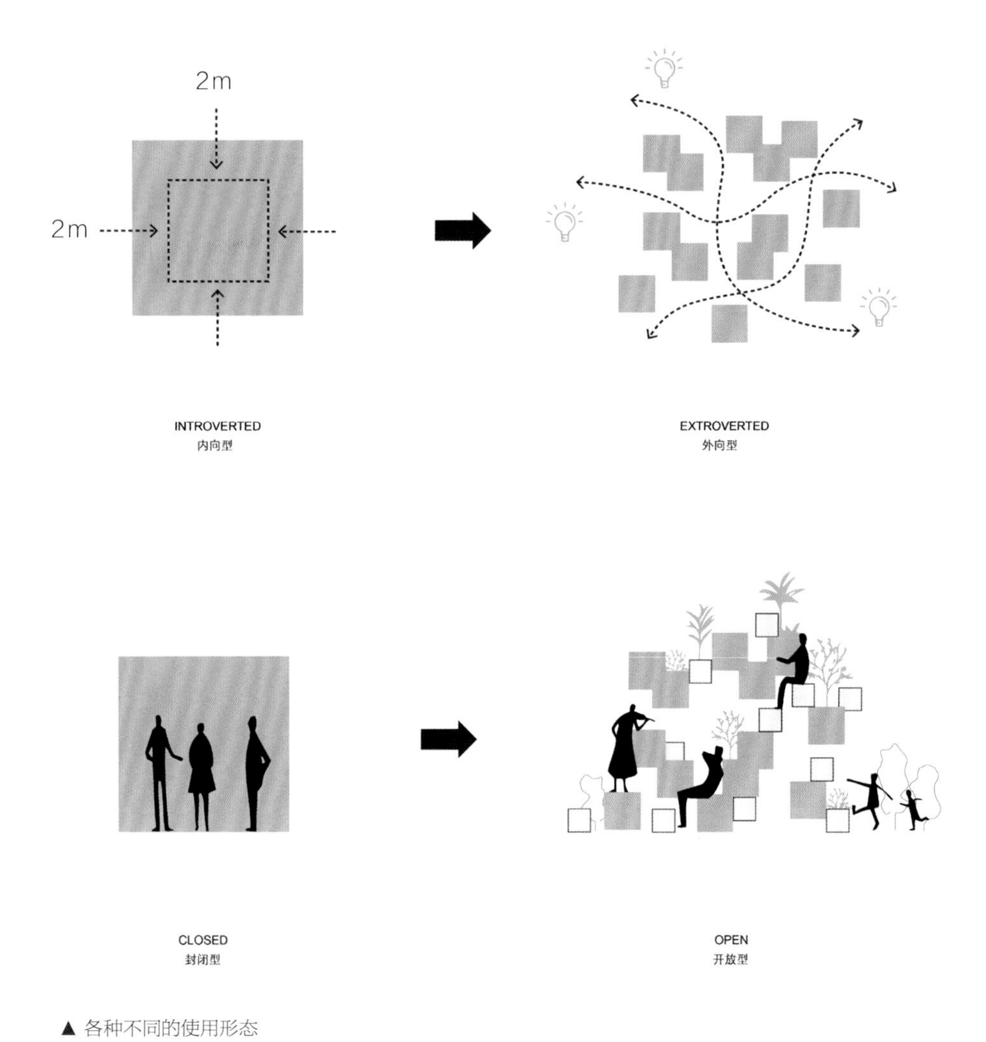

▲ 各种不同的使用形态
Different using shapes

设计师经过大量调研，观察人们在公共社区的活动和行为，尝试创建一个让人们能产生互动的设计，在一天的不同时间支持许多不同的知识创造及分享活动。

After a long period of research, in which the designers observe the activities and behaviors of people in communities, the project aims to create a installation that allows people to interact, and support many different knowledge-creating and sharing activities at different times of the day.

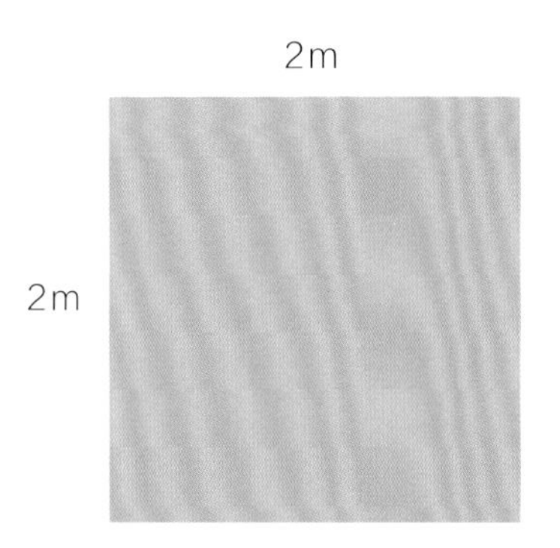

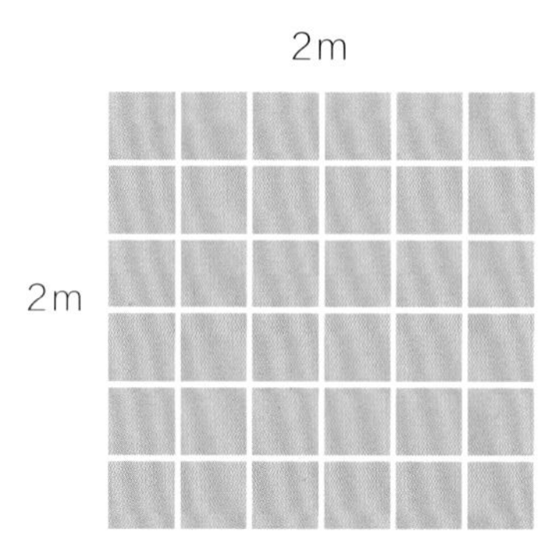

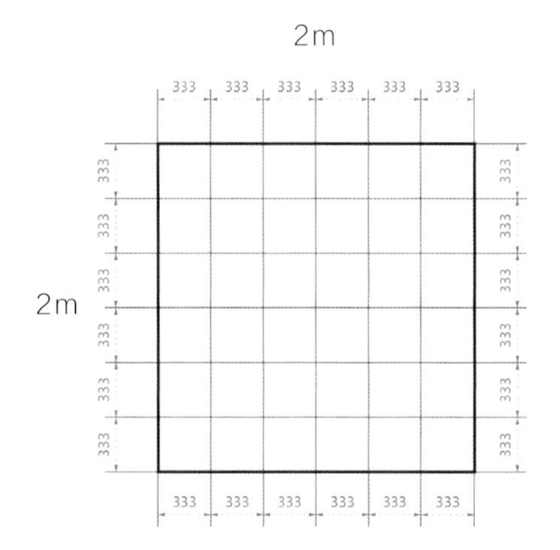

▲ 概念分析图

Concept analysis diagram

2017 年广州设计展闭幕后，这座书架展馆搬至深圳大学继续开放。

Pavilion will be relocated to Shenzhen University and continue its function, after this year's Guangzhou Design Exhibition closed.

▲ 概念分析图

Concept analysis diagram

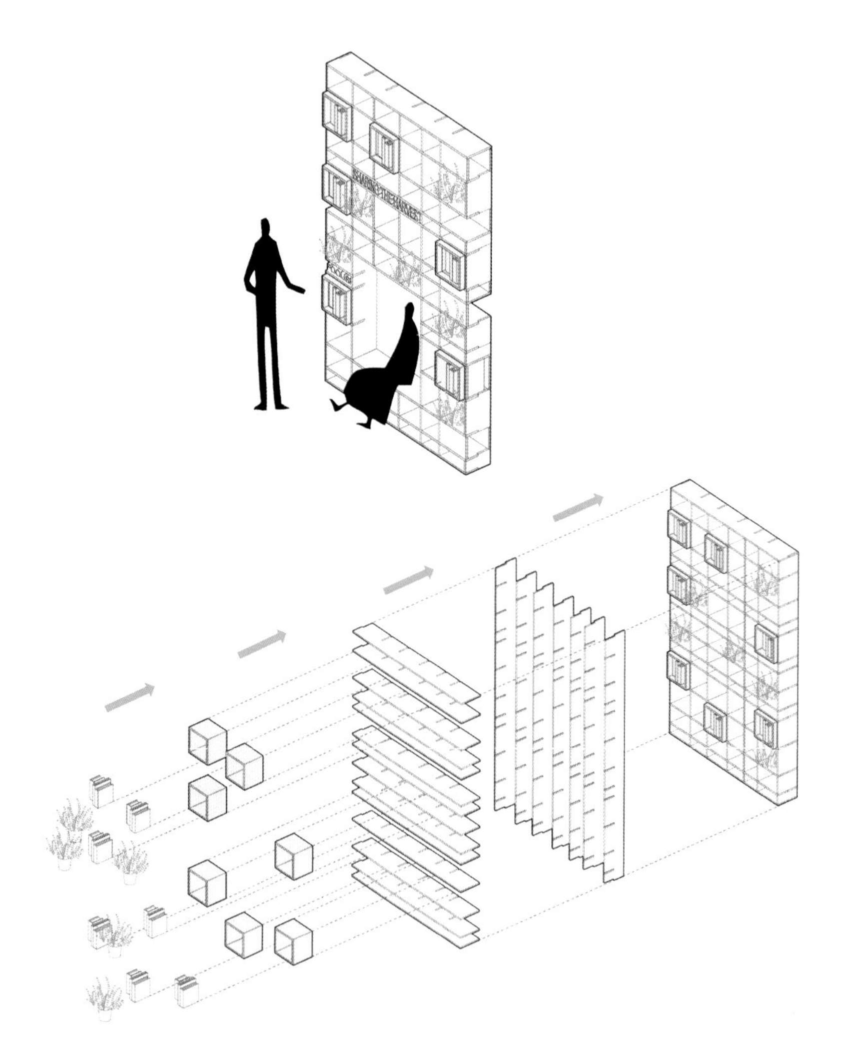

▲ 概念分析图
Concept analysis diagram

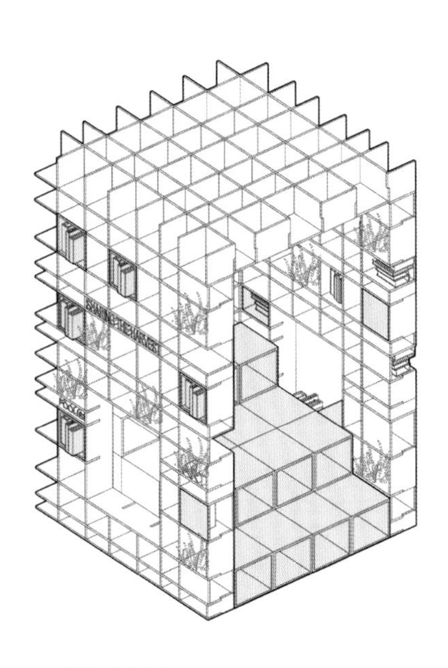

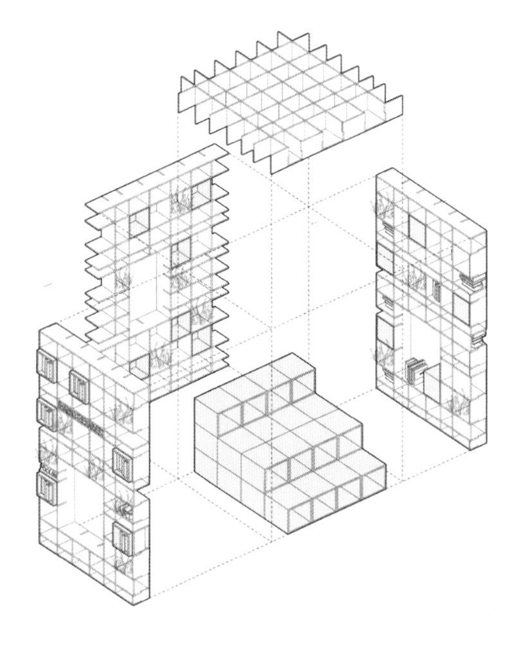

▲ 概念分析图
Concept analysis diagram

整个书架的展出藏品约为 200 本书，它们来自世界各地各个领域的捐献，包括加拿大建筑师 William 、意大利 UD 建筑事务所联合创始人 Momo Li、RUYA 设计工作室 Viva He、插画设计师 CHAN POPO、服装设计师 Sherry Xing、龙瑞律师事务所 Jason、儿童教育心理咨询师 Miao Tang，都捐出了他们认为可供分享的书籍，每一个普通的市民都可以随时发起有意义的知识分享活动。

The shelf displays a collection of approximately 200 books, which is recommended and kindly donated by practicing designers from all around the world , including Canadian architect William, the co-founder of Italian UD architects Momo Li, Viva He from RUYA design studio, illustrator CHAN POPO, fashion designer Sherry Xing, Jason from Long Rui LLP and children's education counselor Miao Tang. Our aim is to encourage every citizen to engage in meaningful knowledge sharing activities at any time convenient for them.

Unit X 拥有数百本书籍和不同的互动装置，不定期的更换为 Unit X 带来新的活动，如读书、演讲、交流会，插花、绘画、小型音乐会、电影放映、民族手工艺展示等。

Unit X contains hundreds of books and different interactive devices. The occasional replacements of the books and devices will bring new action possibilities to Unit X, such as reading, speaking, exchanging meetings, flower arranging, painting, small concerts, film screenings, ethnic handicraft exhibitions, etc. says one of the designer.

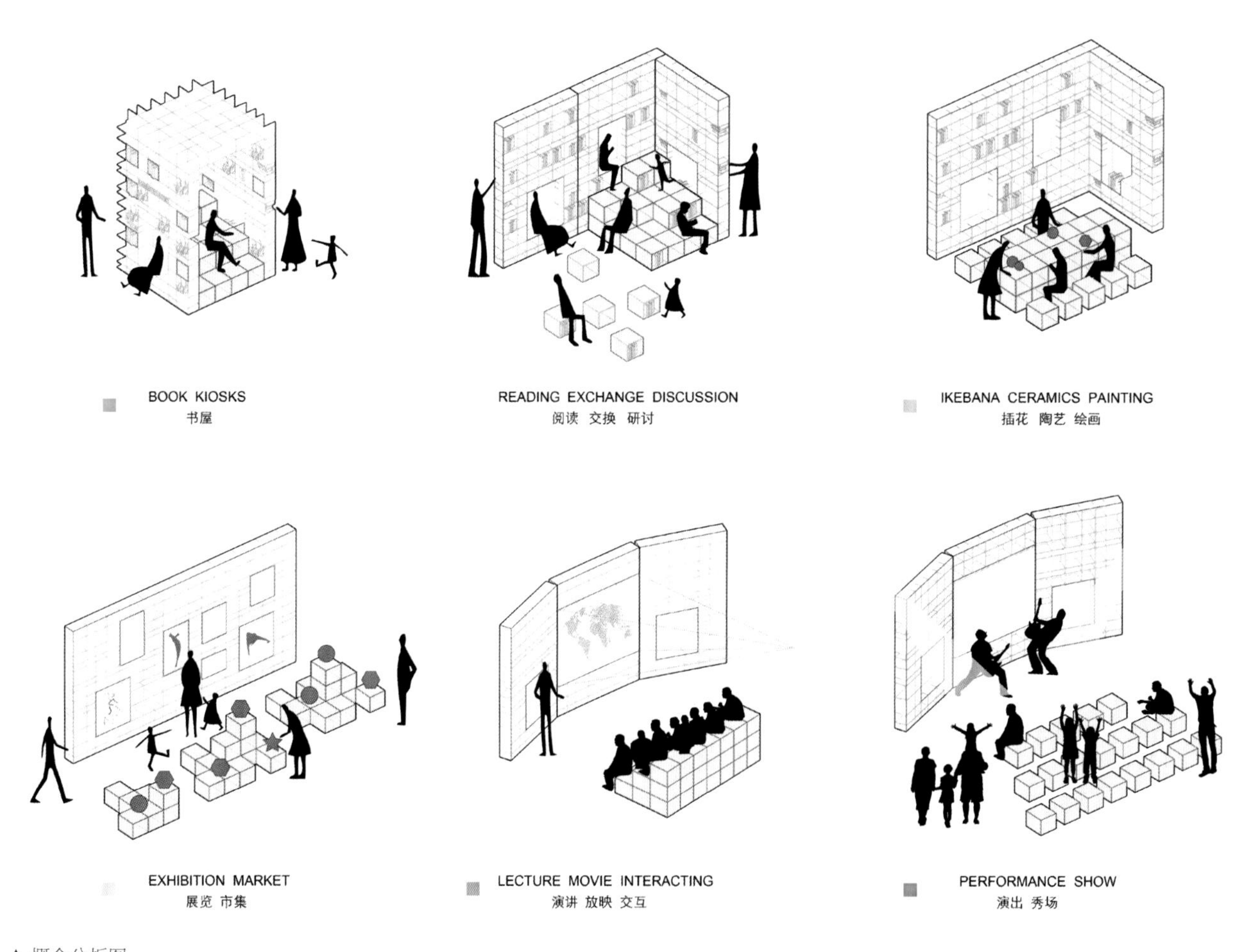

▲ 概念分析图

Concept analysis diagram

▲ 概念分析图
Concept analysis diagram

在孙浩晨和张雷看来，知识共享空间的存在，是超越固定且单纯的生存方式的必要选择，为人们提供有意义的生活条件的一种方式。

在此次展览中，除了作为基本阅读空间之外，这里还将举办各种论坛、讲座，2017 年 11 月 28 日至 11 月 30 日，这个装置在广州设计展展出后，搬至深圳大学继续开放。

▲ 效果图
Design sketch

According to designers Haochen Sun and Lei Zhang, the existence of knowledge-sharing spaces gives us a way to go beyond the fixed and simple living style that merely focus on necessities. It provides people with meaningful living conditions.

In this exhibition, in addition to being the basic reading space, various forums and seminars will also be held. From November 28th to November 30th, this device will serve as a temporary knowledge sharing hub as well as providing a platform on which topics about contemporary knowledge sharing will be dicussed.

Pavilion will be relocated to Shenzhen University and continue its function, after this year's Guangzhou Design Exhibition closed.

▲ 效果图
Design sketch

公园光亭

Park light pavilion

王昊 Hao Wang（中国）

超扁平时代到来，人类视觉体验的二维世界达到一个极限。人们在享受娱乐性与观赏性并存的多元文化的同时，却伴随着个人思想的狭隘与通俗文化的泛滥。

如何跳脱“超扁平”的思维限制，唯有以“超广度”的延伸方式重新去丰富人的所有感官。

在公园中放置一些轻量、半透明的具有治愈作用的共享凉亭。无限单体复制的木制构造与轻质的半透明白色金属铝板网给人一种亲和的接触感。

白天，共享自然。

白天，人可以坐在其中短暂休息，半透明的材质不会完全切断人与自然的互动关系，也不会使人完全暴露在外界。休息者们可以在其中共享短暂的休憩时光，公园中的光、影、风，将会是最佳的共享对象，给人带来无穷的想象空间。

夜晚，共享光源。

晚上，光亭本身会发出柔和的光芒，成为公园中的点状光源，吸引人们环绕过来。人们在有光的光亭附近自发地进行一些活动，个人的感官也会在安静的夜间公园里无尽放大。

With the dawn of the "super flat" era, the two-dimensional human visual perception has reached a limit. Our currently multiple-dimensional culture, albeit entertaining and ornamental, has led to the narrowness of individual thoughts and the dominance of popular culture.

So, how do we get rid of the limitation of a“super flat" mind-set? The only way is to re-enrich all the human senses by extending a“super – wide" mindset.

The light pavilion seeks to "implant" lightweight, translucent and therapeutic pavilions in the park. The multiplication of the wooden structure and lightweight translucent aluminum mesh will bring a sense of affinity and intimacy.

Daytime - sharing nature.

During daytime, people can sit in the pavilion for a short break. The translucent material will not completely cut off the interaction between people and nature. At the same time, it will not make people completely exposed to the outside world.

The natural light, wind and sounds in the park give people the endless possibilities of imagination.

Night – sharing light.

In the evening, the pavilion emits the soft light, which will become the point of light source in the park. This source of light attracts people to gather around it, allowing people to hold activities around the pavilion. It is particularly enjoyable as the quiet night in the park magnifies all senses.

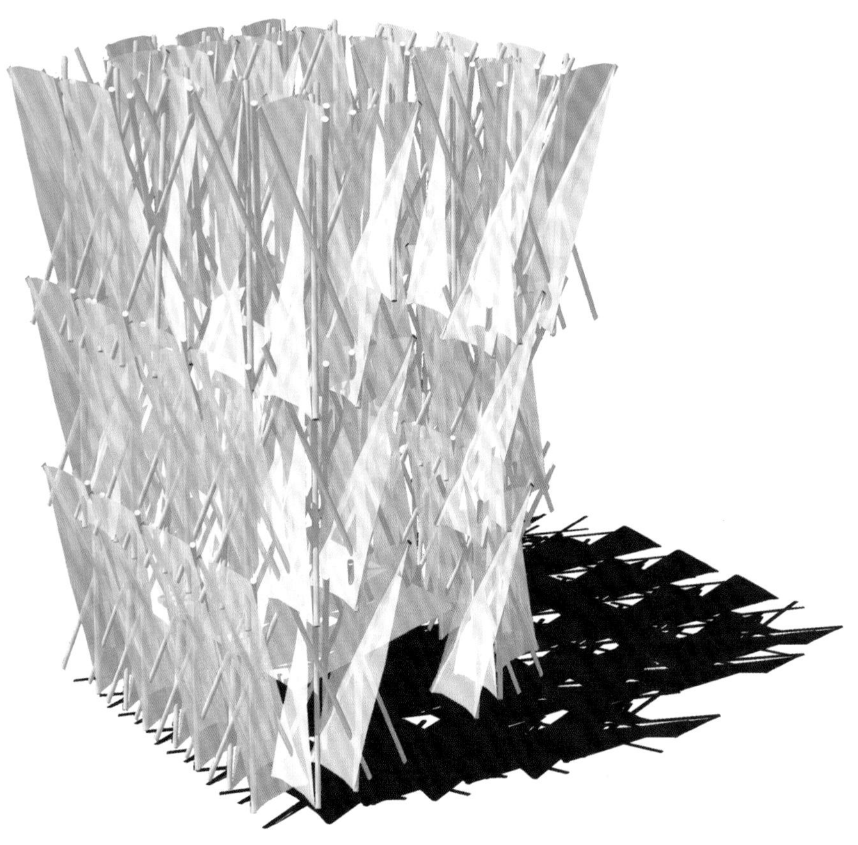

▲《公元光亭》效果图
Design sketch of Park light pavilion

结构：3D 打印接口，高强度 PC 材质。

以组装的方式完成结构的设计，组件可以方便地进行组装与回收利用。

Structure: 3D printing interface and high-strength PC.

The assembly completes the structural design which makes it easy to reassemble and recycle the components.

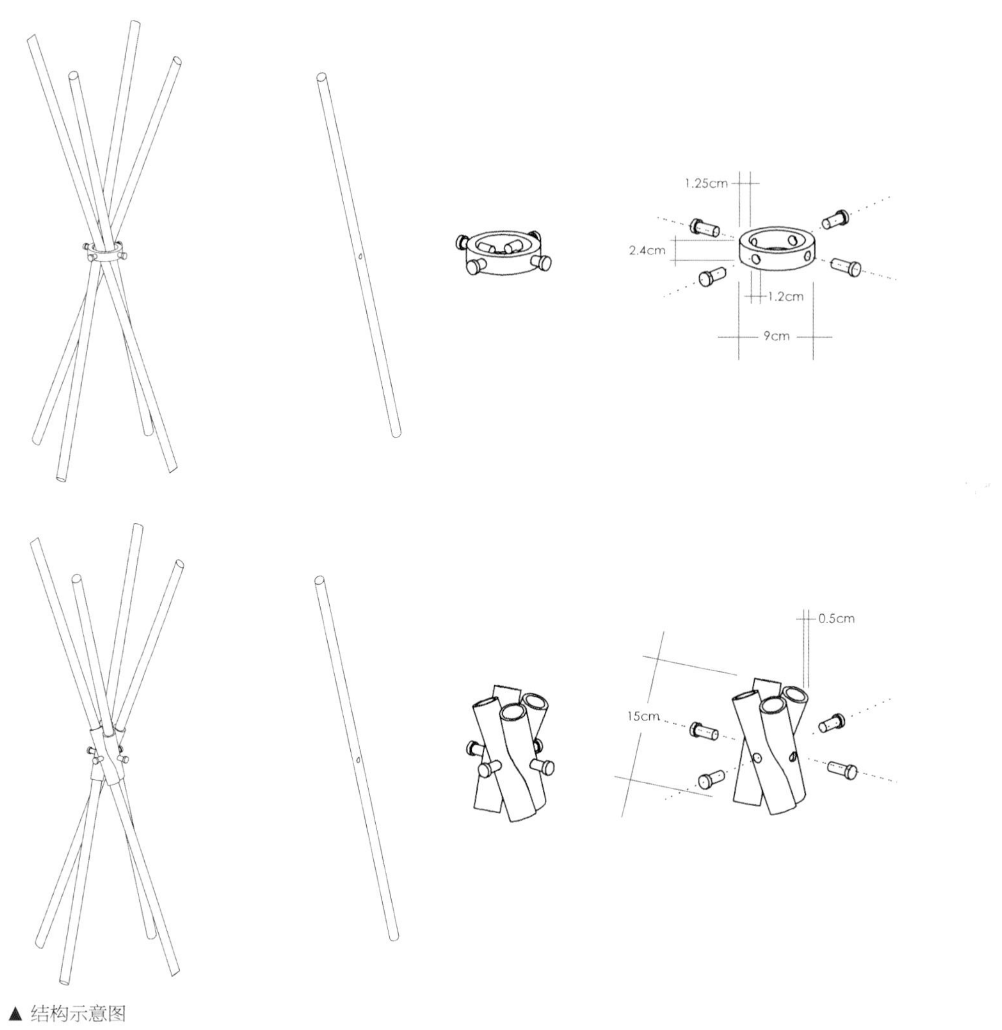

▲ 结构示意图
Structure schematic diagram

材料：白色金属铝板网。

轻薄高密度的白色金属铝板网，在光线的照射下，可以营造整体的“轻质量、半透明感”。较柔软的金属特性也便于塑造自由的弧状曲面，给使用者一种柔和的包围感。

Material: White aluminum mesh.

A thin, high-density and white aluminum mesh creates the "light and translucent" impression with light. The malleable metallic features also facilitate the creation of a free arcuate surface, giving the user a surrounded gentle feeling.

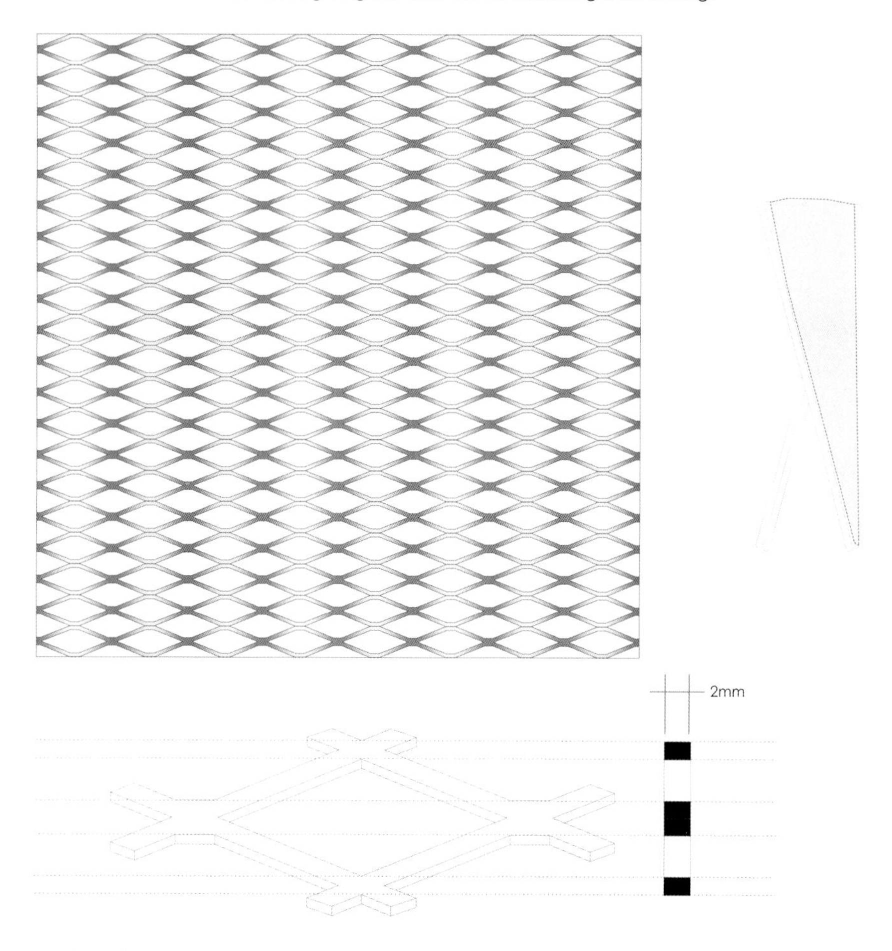

▲ 结构示意图

Structure schematic diagram

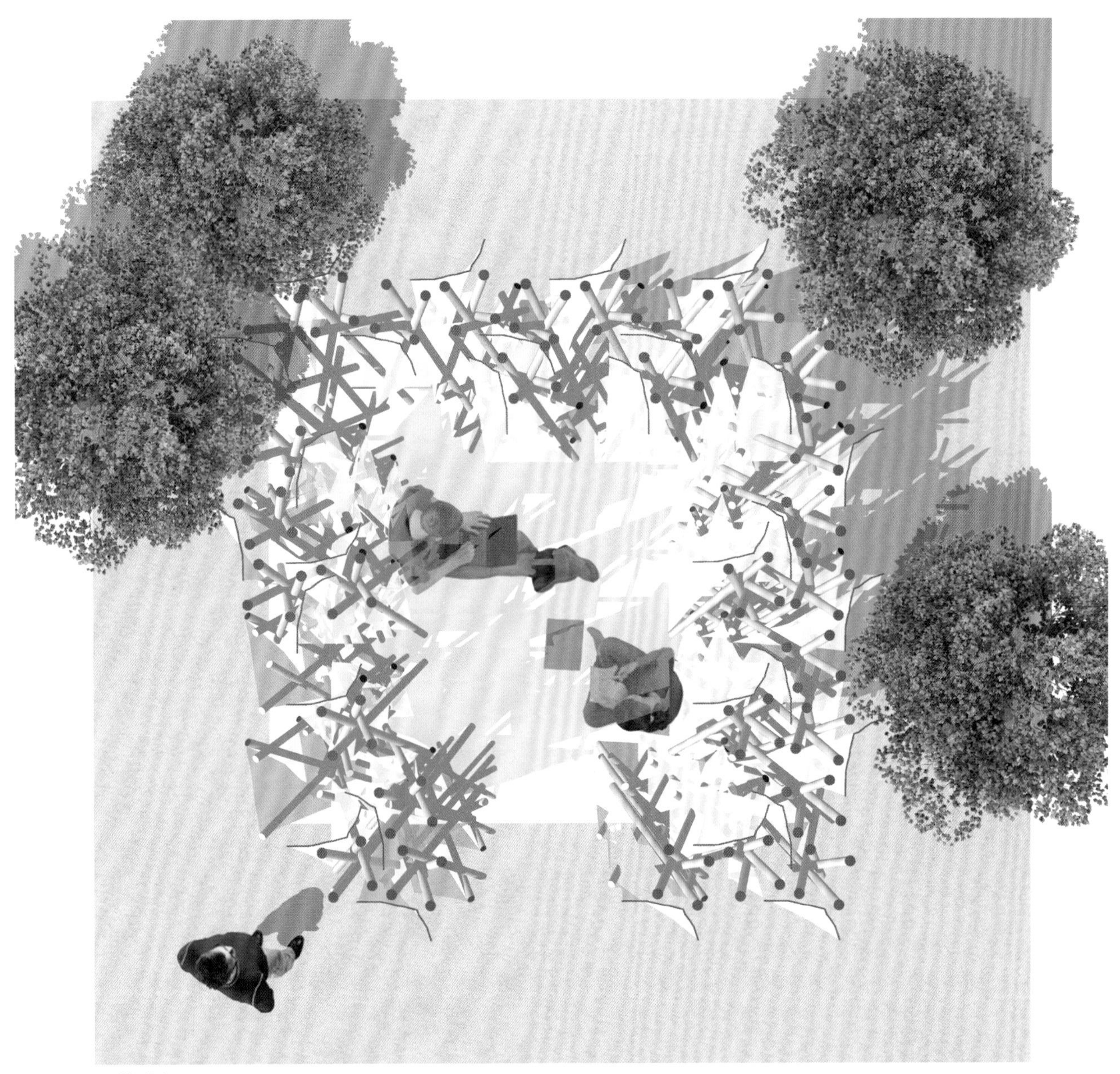

▲ 效果图

Design sketch

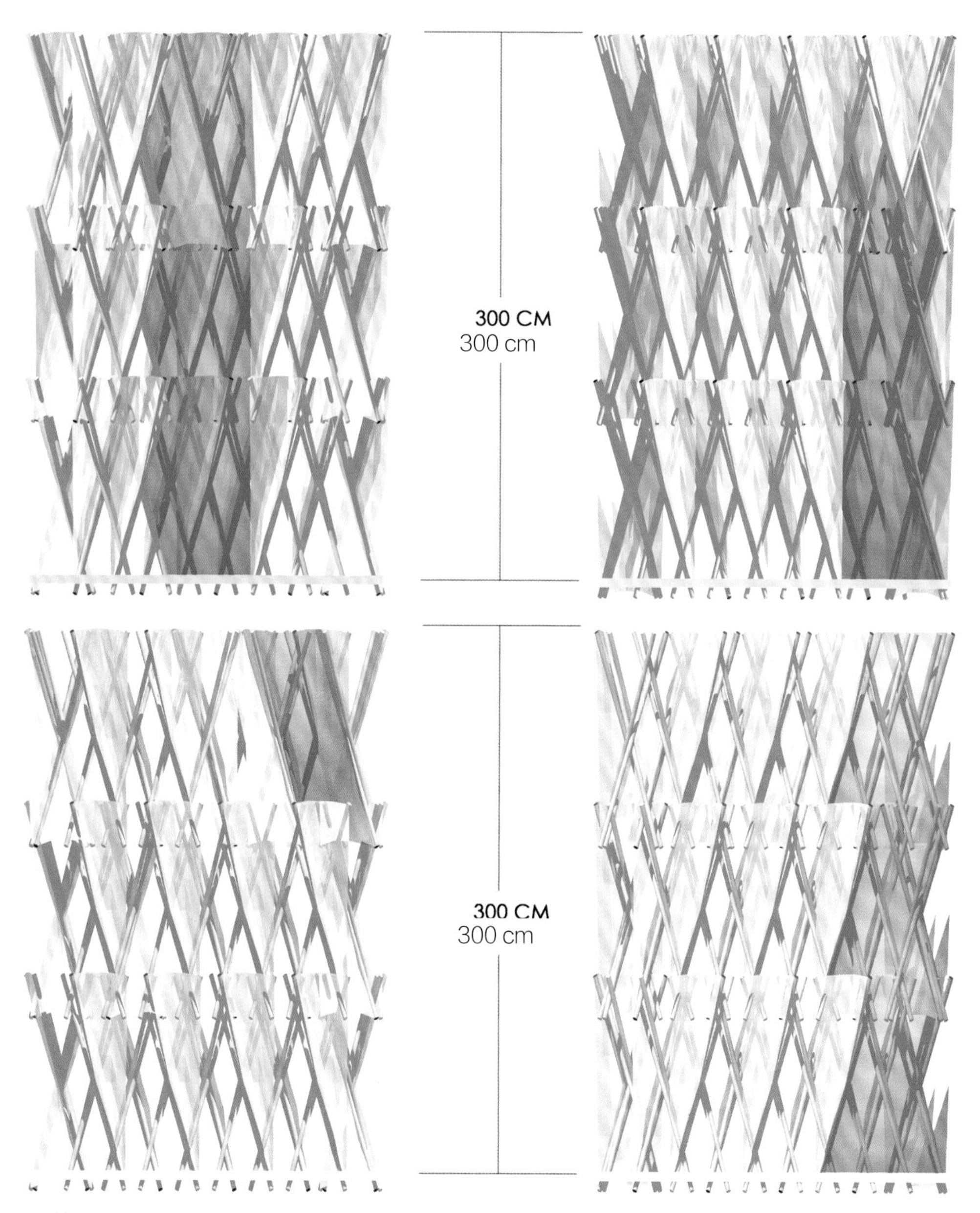

▲ 效果图

Design sketch

▲ 效果图
Design sketch

▲ 效果图
Design sketch

发掘
UNEARTH

张皓翔 Jack Chang（中国）, 张绩 Joshua Teo（马来西亚）

如今，信息的分享速度是超乎想象的。前所未有的信息传播速度让“主流文化”逐渐形成且完全占据了我们的世界。

传统文化慢慢消逝在大众的视线中，近十年来取而代之的是流行文化，似乎要将传统文化埋没了。

传统、本土文化已经沦为次文化，有人想保留仅剩的历史遗迹却没有足够的资金建立博物馆。因此，我们想要借由一个装置让传统文化可以传承下来。那要怎么用一个简单、经济有效的方式恢复、发掘渐渐被遗忘、被埋没的文化呢？

一个博物馆的功能是通过实体和视觉信息传授知识和教育人们，以便我们了解其精神。但同时现代社会信息爆炸，人们对于这些知识已经渐渐麻木。我们想要借由本装置结合广告牌的功能，让信息可以很直接地传达给观众。我们的提案是在城市中设计一系列吸睛的展示装置，形成一个“都市尺度”的博物馆。我们想让这个装置可以跨界、跨城市分享，因此用一个城市的尺度来看这次设计。

In this super flat era, when information is shared at an unprecedented exponential rate, “mainstream" culture has taken over the world by storm.

In recent decade, the traditional culture has been faded away and is gradually being replaced by the pop culture.

These indigenous cultures have been developed into the subculture, which is not well-funded to be preserved by building up a museum. Therefore, we need an apparatus to retain the culture. But, how can we unearth the “forgotten” culture easily and economically?

The purpose of a museum is to educate people through physical and visual information. However, we are overwhelmed by mind-numbing information, and a billboard is illuminating. We propose to create a series of eye-catching installations figuratively constituting a urban-scale museum, which is trans-regional and -urban as well.

▲《发掘》效果图
Design sketch of UNEARTH

我们可不可以将博物馆切片?

Museum:
Can we decompose the museum?

一个解构的博物馆，散播在城市的各角落?

City Map:
Can a decomposed museum be widespread in a city?

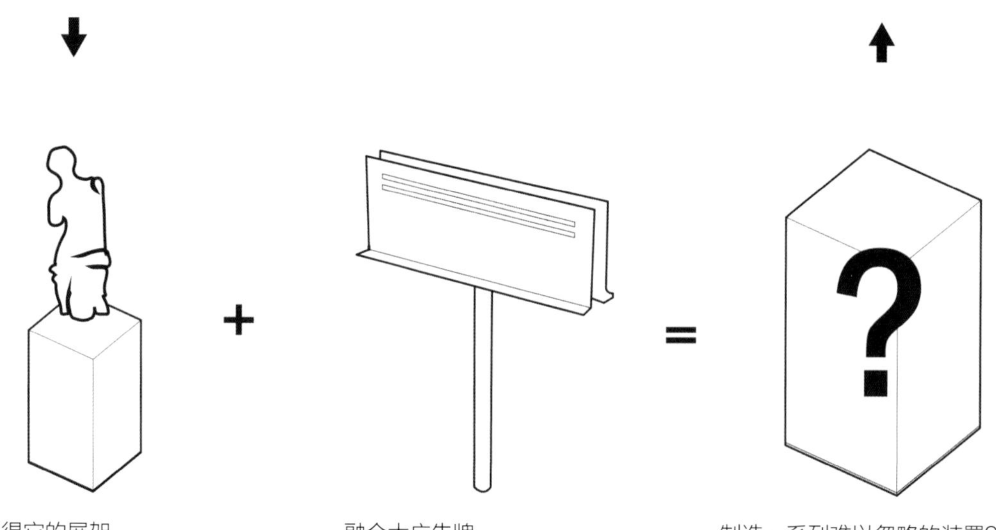

取得它的展架。

Taking its exhibition stand.

融合大广告牌。

Combine it with a billboard.

制造一系列难以忽略的装置?

To create a series of remarkable installations?

▲ 作品创意过程思维导图
Mind Map of work’s creative process

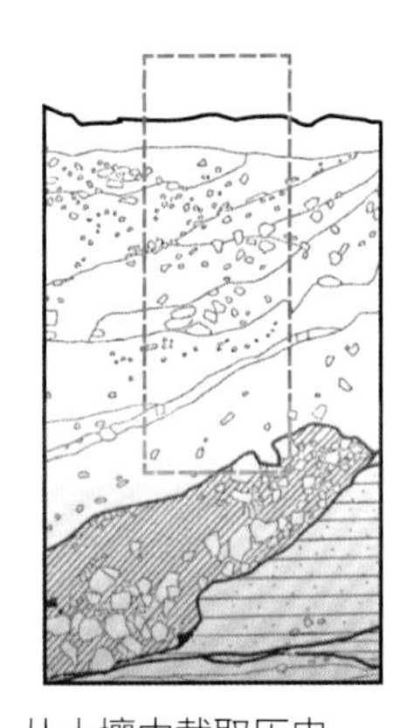

从土壤中截取历史。

Concept:
Slice a piece of the earth.

1.4 m^2 的基地。

1. 4 sq m site.

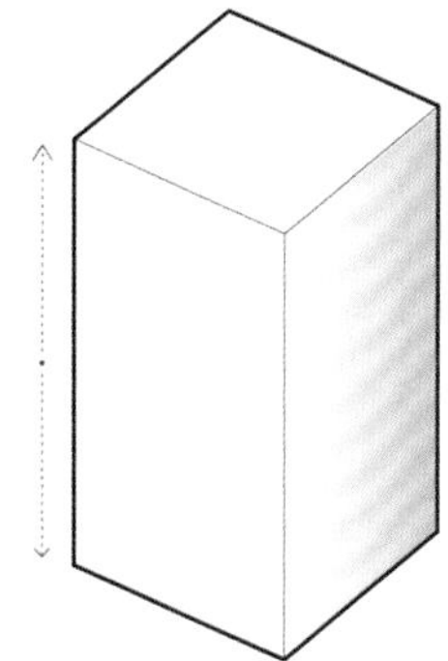

2. 加高 4 m。

2. Heighten by 4 meters.

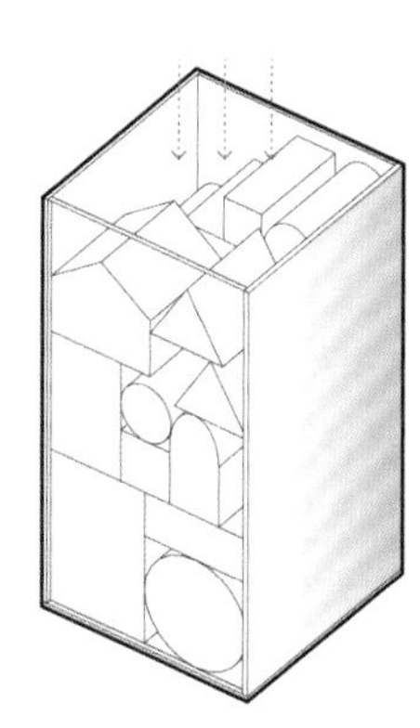

3. 加入展示块体。

3. Add cultural display boxes.

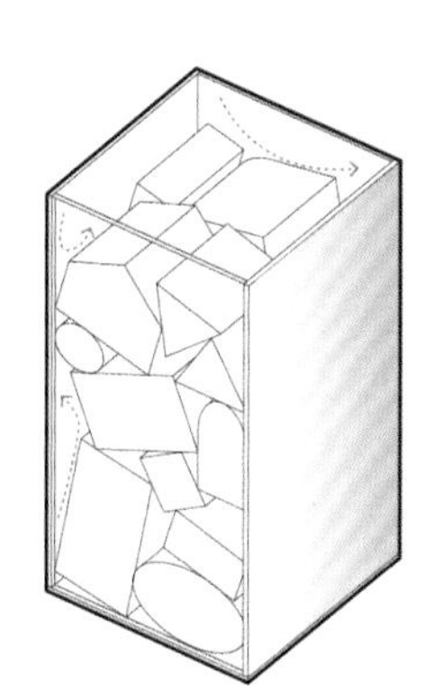

4. 加以施压。

4. Add pressure to boxes.

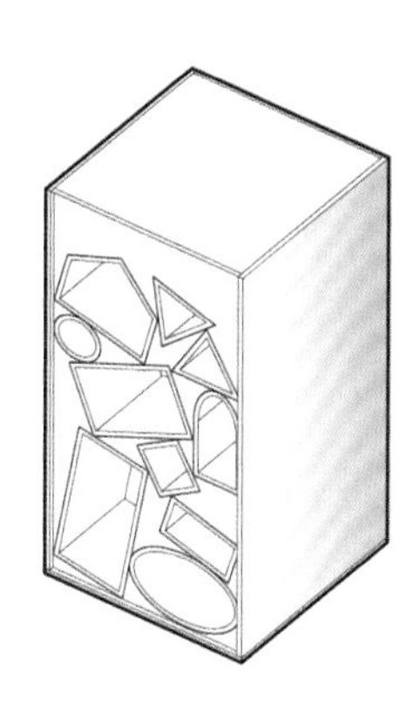

5. 倒入混凝土。

5. Pour concrete into mold.

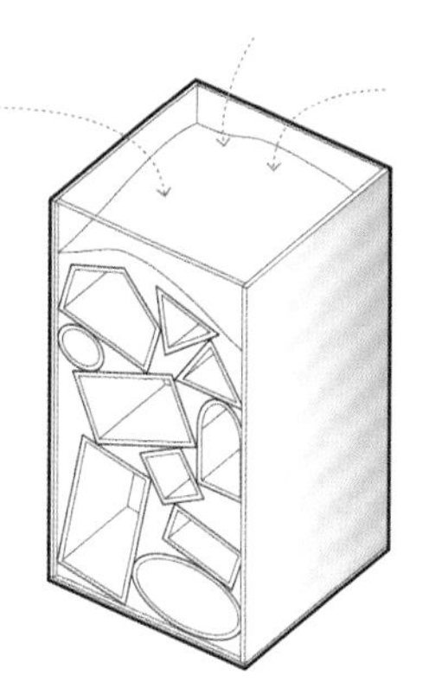

6. 模仿地形。

6. Mimic topography.

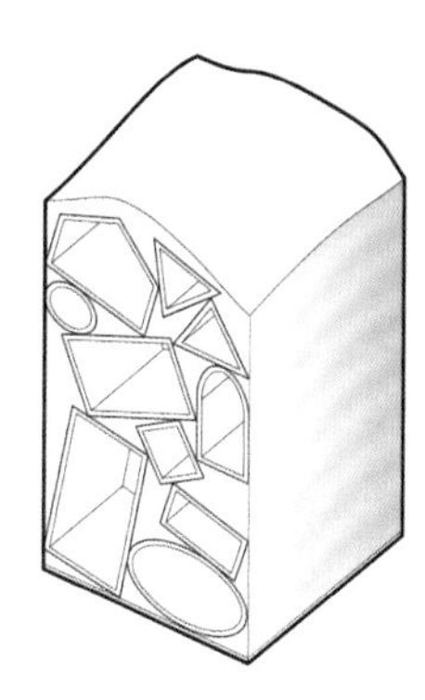

7. 拆模的动作似乎在发掘文化。

7. Remove cast to unearth culture.

▲ 作品创意过程思维导图

Mind Map of work’s creative process

▲ 示意图
Schematic diagram

希望我们的装置不单只是个展品，而更像是一个纪念碑、一个提醒，警示我们不要再轻易遗忘那些珍贵的传统文化。

We hope that our installation is not just an exhibition, but a reminder of the cultures that we have forgotten.

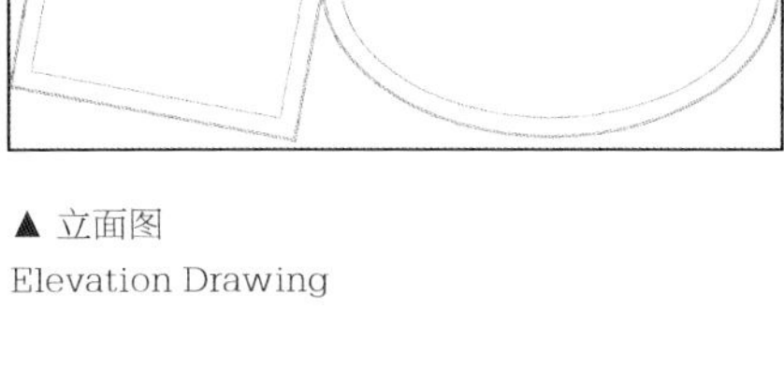

▲ 立面图
Elevation Drawing

▲ 效果图

Design sketch

▲ 效果图

Design sketch

共伴效应
EACH OTHER

闵莹如 Ivy Min（中国）， 陈又宁 Annie Chen（中国）

共享经济现今十分盛行，然而在许多打着“共享经济”名号的商机下，其实存在许多问题，其中共享宠物虽然给予了一些想养宠物却无法实现的人短暂的机会，但却产生了把生命体当作商品交换的道德问题。是否有其他更好的方法呢?

我们很多人都曾经见到过流浪的猫狗。然而很少人会思考这些动物是怎么流落街头的，它们需不需要帮助? 调查显示，大多数流浪动物都是人类弃养所致。

高度灵活的互联网，可以使人们利用手机获得实时信息，去帮助真正有需要的流浪动物。

于是我们创造了一个平台——共伴，在空间上可以给流浪动物提供一个栖身之地。这里有宠物粮，保证动物们的温饱，人也能在此处停留与动物们互动，体验养宠物的乐趣。这里同时有倡导功能，倡导“不弃养，就没有流浪”的理念。而在网络平台上，募资购买饲料，或是运用远端控制系统，为共伴里的流浪动物们添加饲料。

The sharing economy is very prevalent. However, under many business opportunities in the name of the sharing economy, there are actually many problems. Is it possible to offer another option for those who love animals but cannot have one, creating a morale issue — trading lives for goods?

Many of us might have seen the stray dogs. However, few of us would think about how these animals ended up in the streets or whether they need help. According to the statistics, most of stray animals are simply caused by abandonment.

In present, highly flexible economic network can make people help the stray animals in need by using the real-time online information from cell phone.

We create a platform called"each-other," which regards space as a home for stray animals and feeds them with foods. People can stay here and interact with animals, the experiences of which is healing, interesting, and educating people the importance of "no abandonment, no stray." The internet can be used as the platform for either raising fund for foods, or feeding stray animal by remote control over our “each-other” platform.

▲《共伴效应》效果图
Design sketch of EACH OTHER

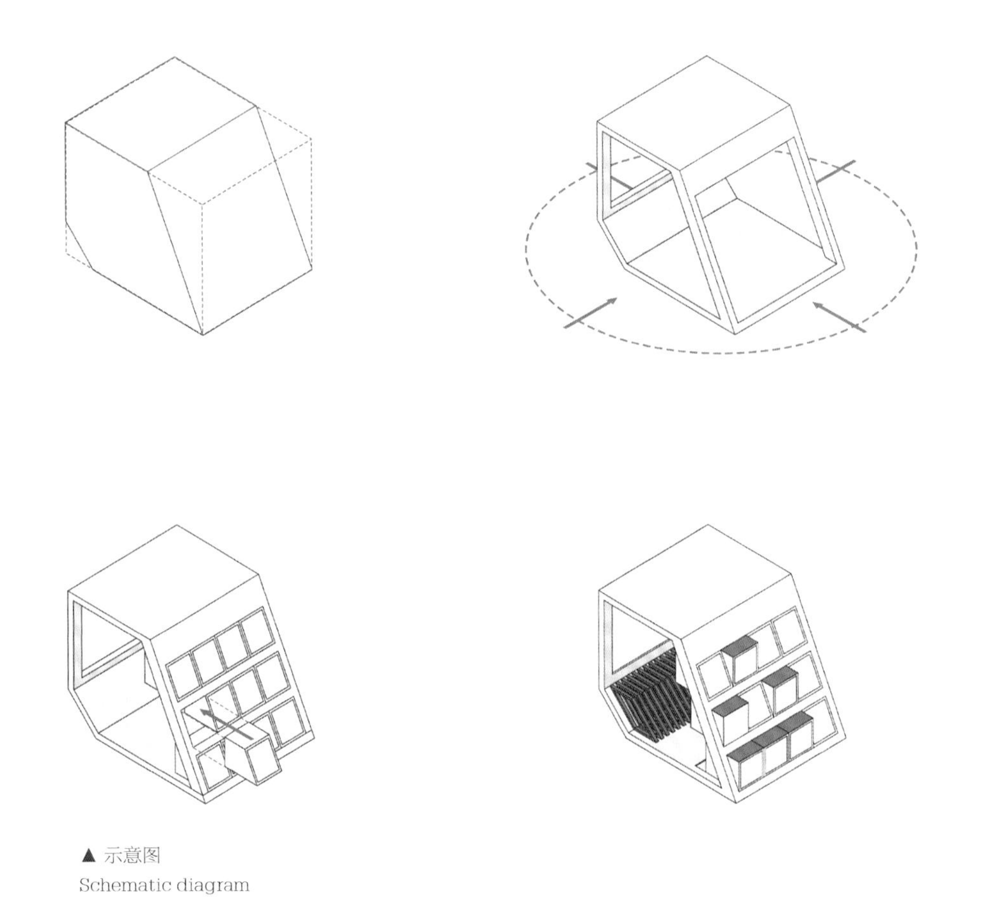

▲ 示意图
Schematic diagram

我们将此空间能够汇聚、交流的元素，转换为一个通透、流动的空间，并且置入能够让人停留的设施，最终呈现出这一空间——共伴。

人们蹲下观看盒内动物的同时，看到背景正播放着动物视角的影片，借此了解流浪动物的悲惨生活，以达到提醒人们“不弃养，就没有流浪”的功能。

We transform the converging, exchanging elements into a transparent, circulative object, which is turned into our design work “each-other.”

While squatting and watching the animals in the box, people can see the animals’ perspective videos played in the background. The videos showing the miserable life of stray animals so as to remind people of "no abandonment, no stray."

▲ 效果图

Design sketch

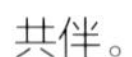

共伴。

Each other.

柜体与动物的关系。

The relationship between the cabinet and animal.

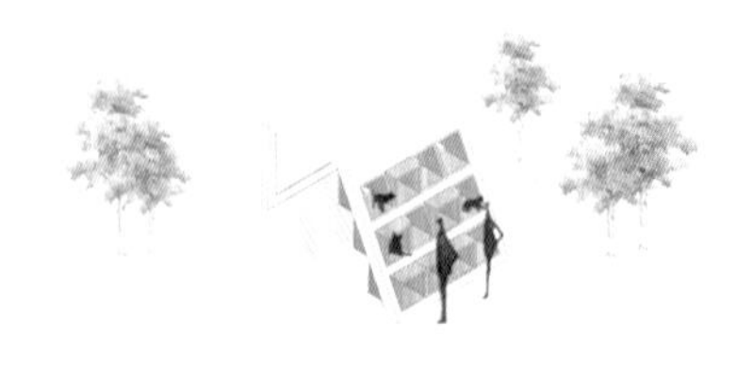

人与动物和视频的视线关系。

The visual relationship among people, animals and video.

使人与动物产生互动。

The cabinet makes people interact with animals.

人与动物共同休憩。

Sharing leisure moment with animals.

因柜体产生人和动物的互动和汇聚。

Because of the cabinet, people and animal assemble and interact with each other.

▲ 概念分析图
Concept analysis diagram

动物胶囊橱窗：参考一般猫狗所需最小空间制成，给予动物短暂的庇护，并且通过网络和电力感应给予饲料。

Capsule window for the stray animals: this product has the minimum required space for cats and dogs, provides the shelter to stray animals, feeds animals by utilizing internet and life-sensing device.

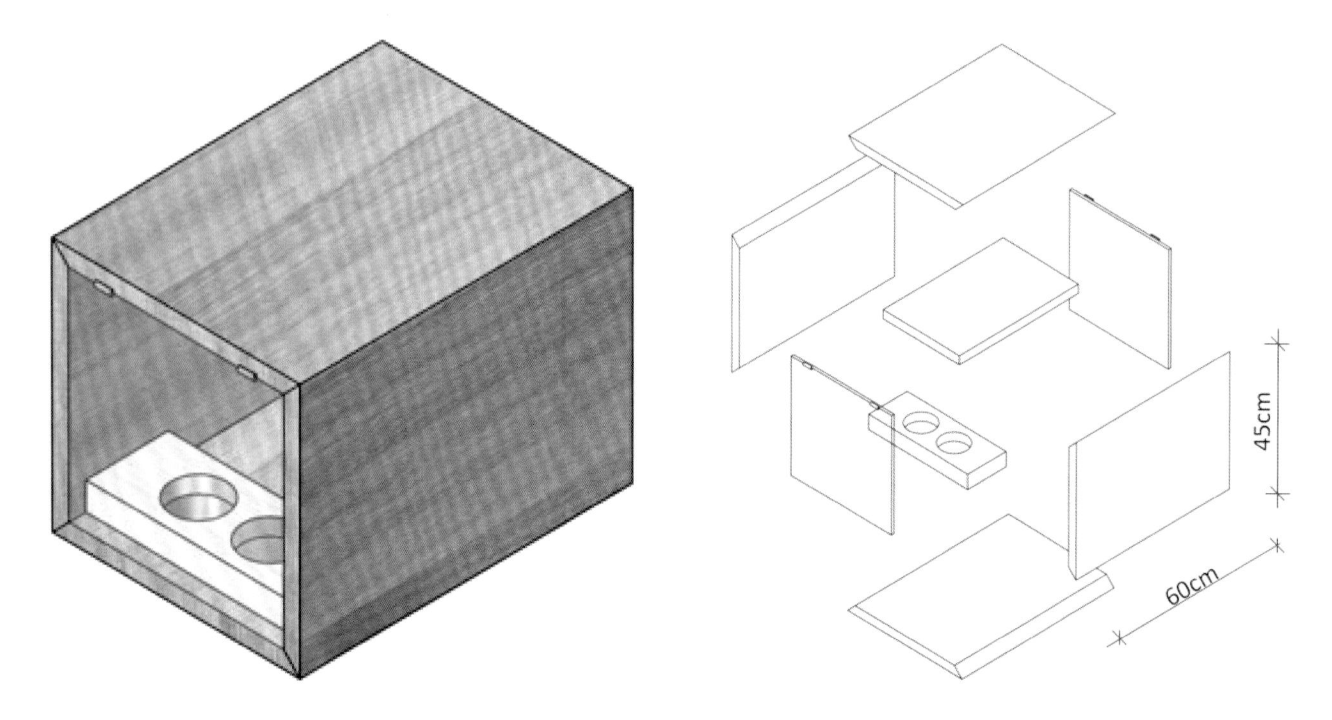

▲ 概念分析图
Concept analysis diagram

望
FINDING: THE HOPE

黄华敏 Min Huang（中国）、高语辰 Eunice Kao（中国）

在这个信息极度扁平化的时代，时空正在快速地被科技挤压，在这个时间与空间迅速压缩和消失的时代中，原本的共享已然变质，现在的共享是自愿与非自愿地分享与被分享，爆炸的信息量被极致地碎片化，压缩并公开。 倘若没有能力去整理消化，我们将为信息所奴役。

我们成了科技时代的潘多拉，打开了盒子，释放出我们知道的、不知道的好处及坏处。其中，希望不是不存在，而是被锁在盒子里，等待着能寻找到它的那个人。

In the "super flat" information era, space and time is aggressively being "flattened" by technology. In the era when time and space can disappear and be compressed, the original meaning of "sharing" has currently changed into voluntarily and involuntarily sharing or being shared. The overwhelming information is being fragmented, compressed and published. If we are unable to organize and comprehend the information at hand, we will become the slaves of information.

As if unlocking the Pandora's Box, we cannot identify either the benefits or advantages. Not absent but being secured in the box, hope is waiting for being uncovered.

▲《望》效果图

Design sketch of FINDING：THE HOPE

当今信息源头方面是多元且没有限制的，往往随着信息的传播，信息中原本的核心思想变得越来越不重要。

当信息的核心不再重要，信息对于我们将会是囫囵吞枣式的大轮廓，还是更加完整紧密的排列重组呢？

In this new era, the source of information is diverse and unrestricted, which magnifies the peripheral scale. This means that the so-called “central idea” will not be as important as ever.

When the core value of information becomes insignificant, will information be something we consume senselessly, or be reorganized firmly and completely?

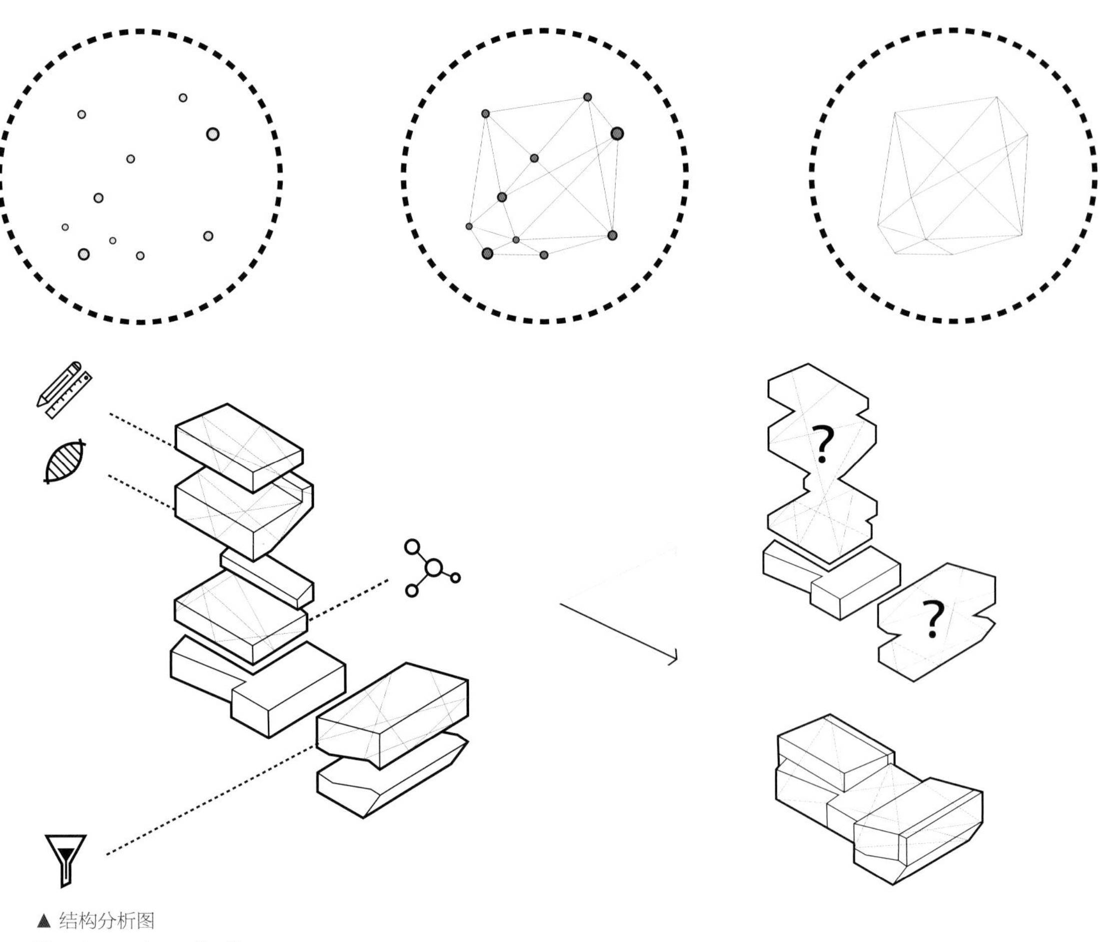

▲ 结构分析图

Structure schematic diagram

▲ 效果图

Design sketch

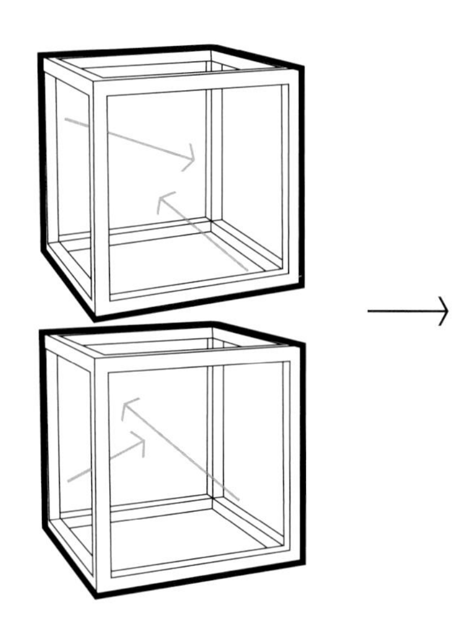

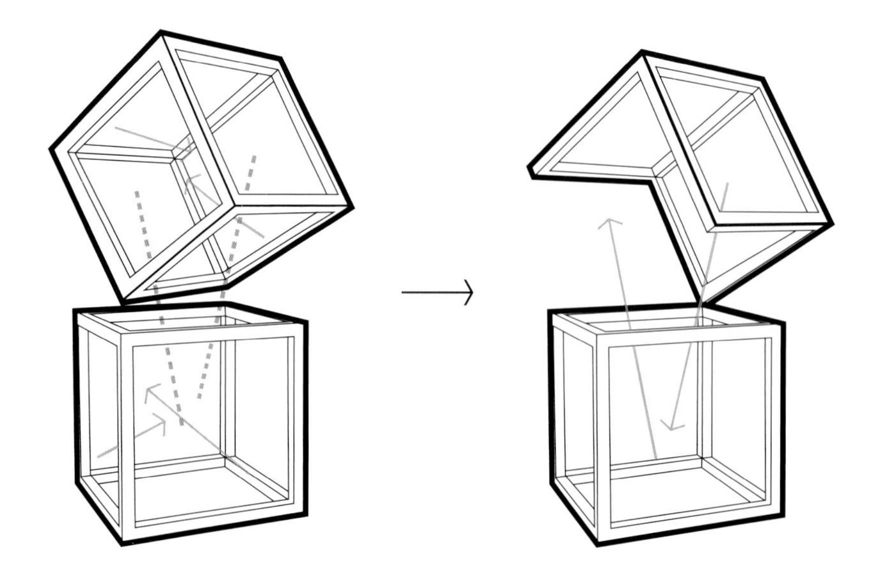

以前的资源共享是可以被控制的。

Initially, resource sharing can be controlled.

扁平化时代来临，共享体制开始动摇。

The “super flat” era is coming, and the sharing system is shaking.

当代的资源已达到全方位的分享与被分享。

At present, resource has been comprehensively sharing and shared.

▲ 概念分析图

Concept analysis diagram

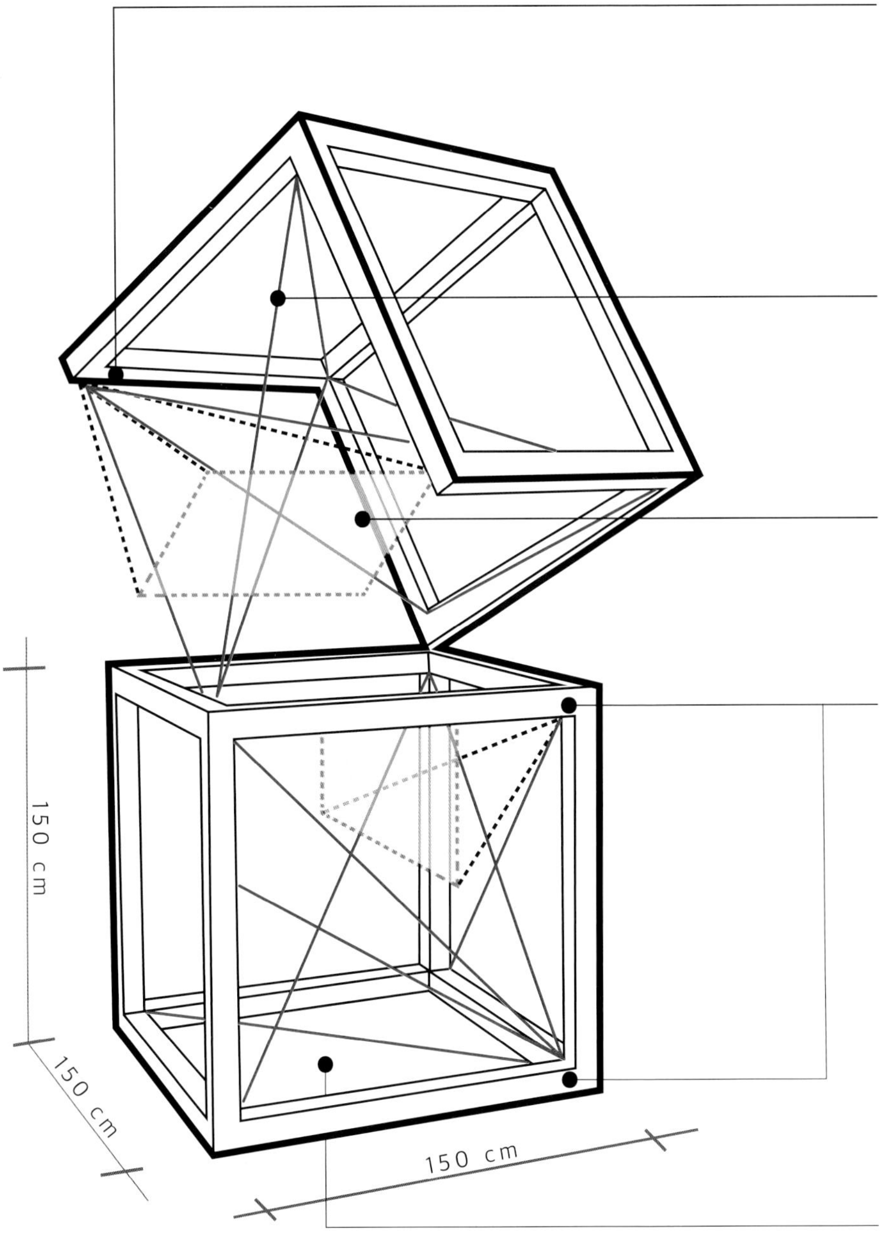

投影光，象征当代的希望。

Light symbolizes the modern hope.

绑线，暗喻当代信息共享的迅速和杂乱。

The tangled lines imply the status of swift and chaotic information sharing.

投影信息，呼应实体信息时代的不再。

The projection of information symbolizes the end of a physical information.

木条框架，强调共享信息看似容易取得，实际想要进一步了解却有困难。

We use a wooden frame to emphasize that sharing information may be easily to obtain, but in fact difficult to further comprehend.

刻意缩小容纳空间，仅能容纳一人，强调个人的选择与找寻在当代共享概念中的重要性。

Deliberately minimizing the space to the one-personal scale, we focus on the personal choice and modern sharing.

▲ 概念分析图

Concept analysis diagram

共感

SHARED HEART

曾子麟 Andy Tsang（中国）

音乐能令人产生共鸣，人类通过不同的音符表达情感。社会上有时候人与人之间缺乏一个沟通方法去打破彼此之间的隔膜， 通过共享音乐平台，人们可以在空间内聆听同一首音乐， 打着相同的拍子， 在细小的 2 m× 2 m 的空间内感受共同的气氛。

装置平台内的墙身把人们的视线阻隔了， 令人们在其中不会觉得没有自己的空间，由此来回应现今社会人们对空间的疑问：“我们需要怎样的空间？ ”设计师希望可以通过这个装置令大众再讨论这个问题：共享空间是否是一个方向？

Music is a striking world language. People express emotions through different notes. In the modern society, people need a better communication skill. Through the shared music platform, people can listen to the same music, play the same tempo, and share the same atmosphere in the 2 m × 2 m space.

The walls in the platform obstruct the view, and people do not feel that they lack personal spaces. It signifies a response to the contemporary question of space, “What space do we need” . The designer hopes that the platform encourages people to discuss the topic “Is the sharing space a direction?”

▲《共感》效果图
Design sketch of SHARED HEART

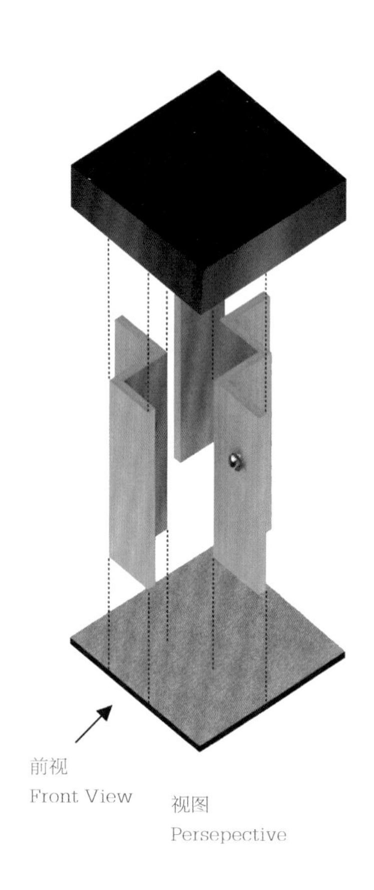

▲ 效果图
Design sketch

前视图
Front View

后视图
Back View

侧视图（左）
Left Side

侧视图（右）
Right Side

音乐可拉近人与人的关系，令我们产生共同的感觉。

Music is world language which makes people feel the same.

我们缺乏空间，令生活变得孤独。

We lack space, making life lonely.

如何令大家聚在一起分享同一种感觉?

How can we make us together share the same feeling?

▲《共感》概念分析图

Concept analysis diagram of SHARED HEART

22×2 的接触

22 × 2 CONNECTING

留鸿运 Elvis Liu（中国）

关于共享：

我们身为不同的文化个体，因为共享而共识。

我们把家里闲置的部分租借给旅客，我们为了参加活动而租用一件礼服，我们共乘一辆车到不同的地方。

这些共同经历的时间与物件，显现了人们行为的一致性。在行为里，拥有相似的喜怒哀愁与面对生活的态度，相似的地方成了我们之间的沟通方式。经过沟通，我们找到共识，被理解的瞬间、情绪释放的瞬间，我们才知道与人接触就是最大的共识。因共享而共识。

我们送礼物，是希望对方了解自己的心意。通过送礼物，我们得以与人交谈，联系感情之余也会获得不同方面的帮助。送礼物的概念一直影响着人们的行为，大家有共识似的，去见朋友时会主动带礼物。而这一行为，体现了共享的核心概念——主动的共识，通过送礼物的行为达成共识，以保证双方关系在未来有良好发展。

About Sharing,

We come from different cultures, and we unite from sharing.

We lease the unoccupied residential space to tourists, we rent and share the same dress, and we share the ride.

The same experience in time and object signifies the uniformed human behavior. In the behavior pattern, we have the similar emotions of happiness, angry, sad and worry, and the attitudes to life. The similarities have become the communication among people. By means of communication, consensuses are founded. Experiencing the moment being accepted and releasing emotions, we learn that human contacts are the biggest consensus. We share, and we unite.

In the eastern tradition, sending gifts implicitly symbolizes one-way communication, which was divergent from the currency society and dated back to the era of gift economy. During the gifting process, people talk, bond and occasionally receive helps. The concept of sending gifts has been influencing human behavior. Visiting friends with gifts has become the consensus, which indicates a core idea “consensus Initiative.” Gifting leads to consensus, and consensus leads to promising mutual interaction.

▲《22×2 的接触》效果图
Design sketch of 22 × 2 CONNECTING

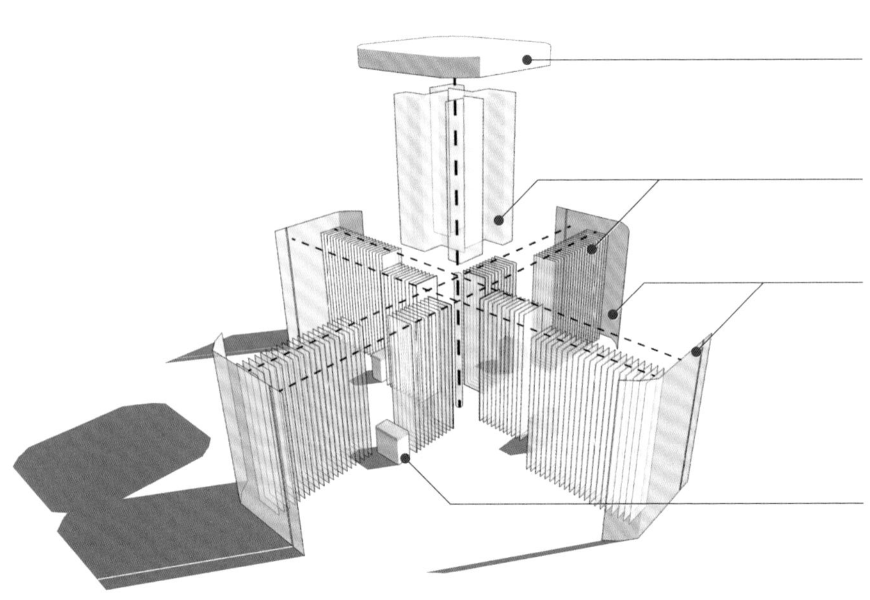

▲ 结构分析图

Structure schematic diagram

22 × 2 指的是 22 对染色体，暗喻生而为人的共通性。在展厅中，利用透明且柔软的纸质材质建成 2 m × 2 m 的展区，柔软的质感对应人的情绪的多样性。而狭窄的走道、 相互接近的座椅，除了提供多处接触的可能性，也寓意在共享时代中的我们，也应该有近距离的接触。

22 × 2 means 22 pairs of Chromosomes, which imply the human commonality. In the exhibition, the soft and transparent materials constituted a 2m × 2m space. The soft texture corresponds to the diversity of human emotion. The narrow aisle and closed seat not only provide many possibilities of contact, but also indicate that we are living in the era of sharing and are contacting firmly with people.

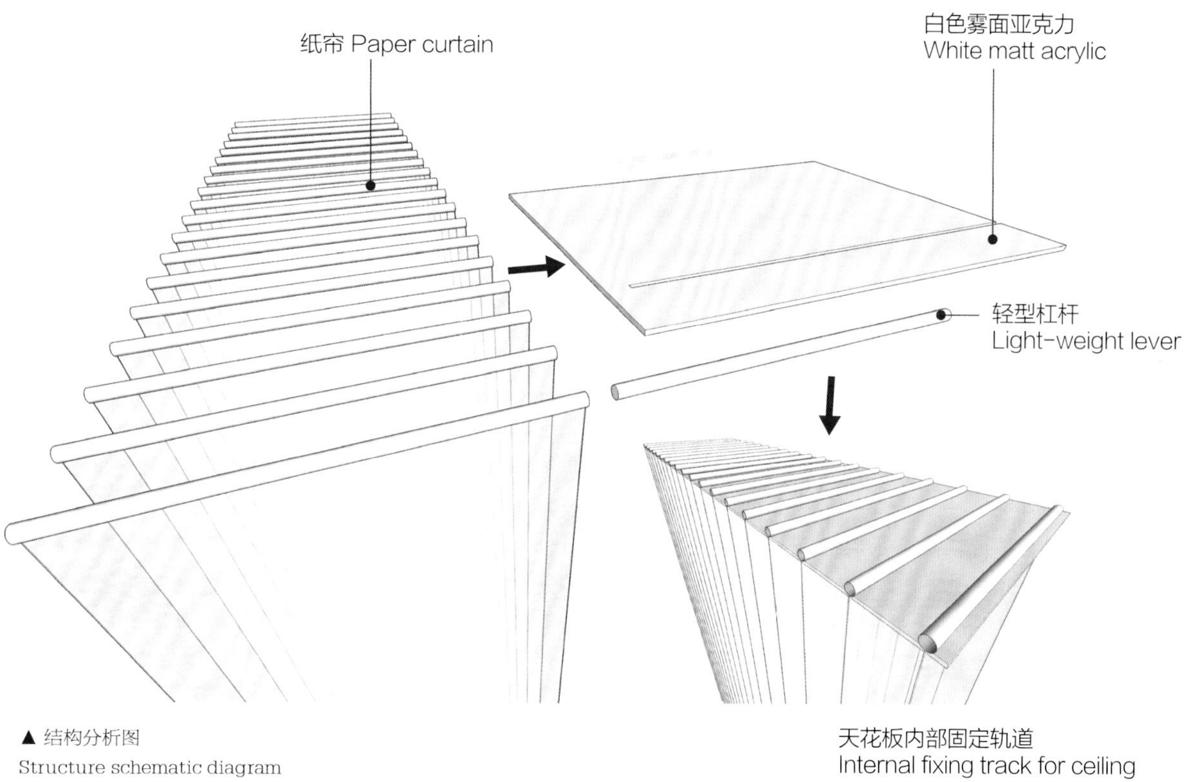

▲ 结构分析图

Structure schematic diagram

触动
TOUCH

施嘉晓 Sison Shi（中国）

社会慢慢转变成共享模式，但现代人生活节奏越来越快，慢慢就会忽略人与人之间最纯粹的交流，本次设计以“触动”为主题，用一根根圆管作为材料，希望可以创造一个短暂的交流共享空间。在圆管中切出非线性的路径，这里没有座椅，只是人们经过的一个中介空间，这里具有模糊性、轻透性、动态性，以虚实相交的手法，让身在其中的人慢慢渗透在这个场景中。

空间虽然相隔，但我们在相隔的管子上开了一个个洞，高、低开口，让身在空间的彼此，有似见非见的感觉，只靠声音与人纯粹地交流，形成一个共享的聊天平台。轻轻地触动一下，便会让彼此坠入感官的情绪中，回归到最纯粹的交谊。

Though the sharing society is slowly developing, the fast-pacing modern life sinks the instinct human communication into oblivion. Named “Touch,” the design work composed of circular tubes creates a transitory space for sharing and communicating. Given the nonlinear path created by the round tubes, there were no seats, and are accessible only for people walking by. The fuzzy, light, transparent, dynamic and flexible design skills created sensational experiences to people as well.

The holes on the curved tubes imaginatively facilitated communications by voice transmission, which inherently established a communication platform. The gentle sensational vibrations transformed the emotions into the most primitive status.

▲《触动》效果图

Design sketch of TOUCH

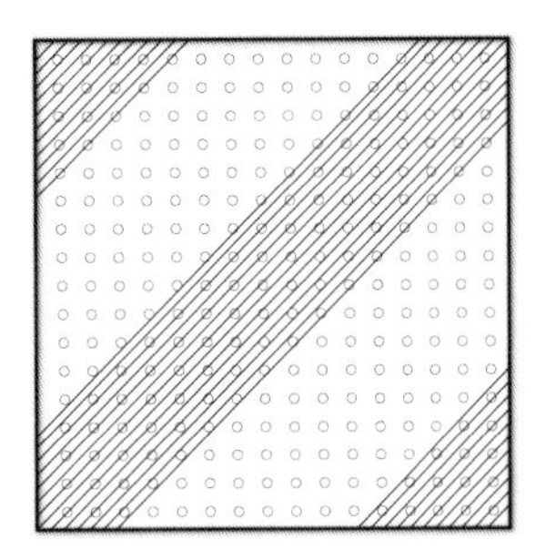

利用圆筒排列，再用路径去切割出动线。

The cylinders were arranged into routes.

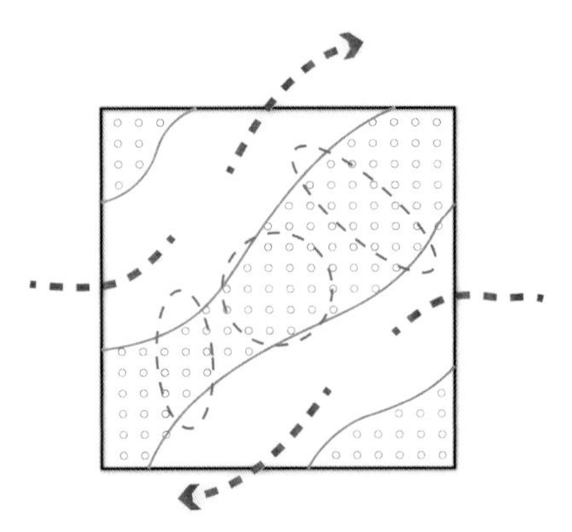

利用曲线分割，创造出两条相对的路径。

The crooked dissections created two comparable space experiences.

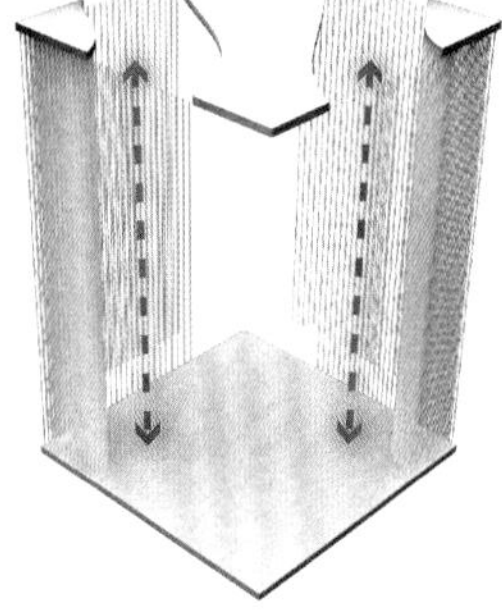

垂直方向的天窗，身处其中，可以与自己内心对话。

The upper windows artistically signified the inner conversation.

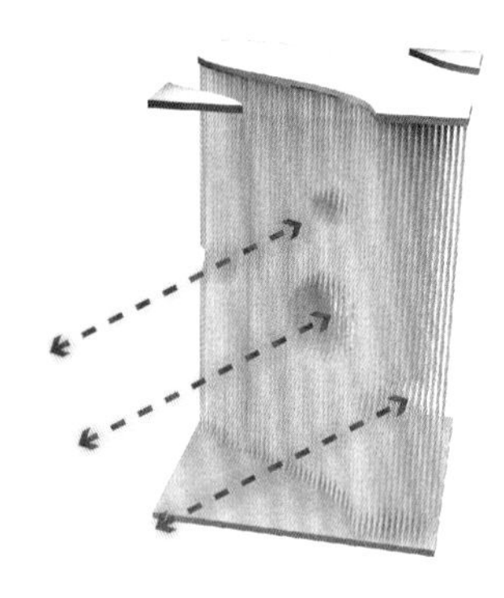

水平方向的开口，模糊的互动。

Horizontal openings manifested the obscure interaction

▲ 概念分析图

Concept analysis diagram

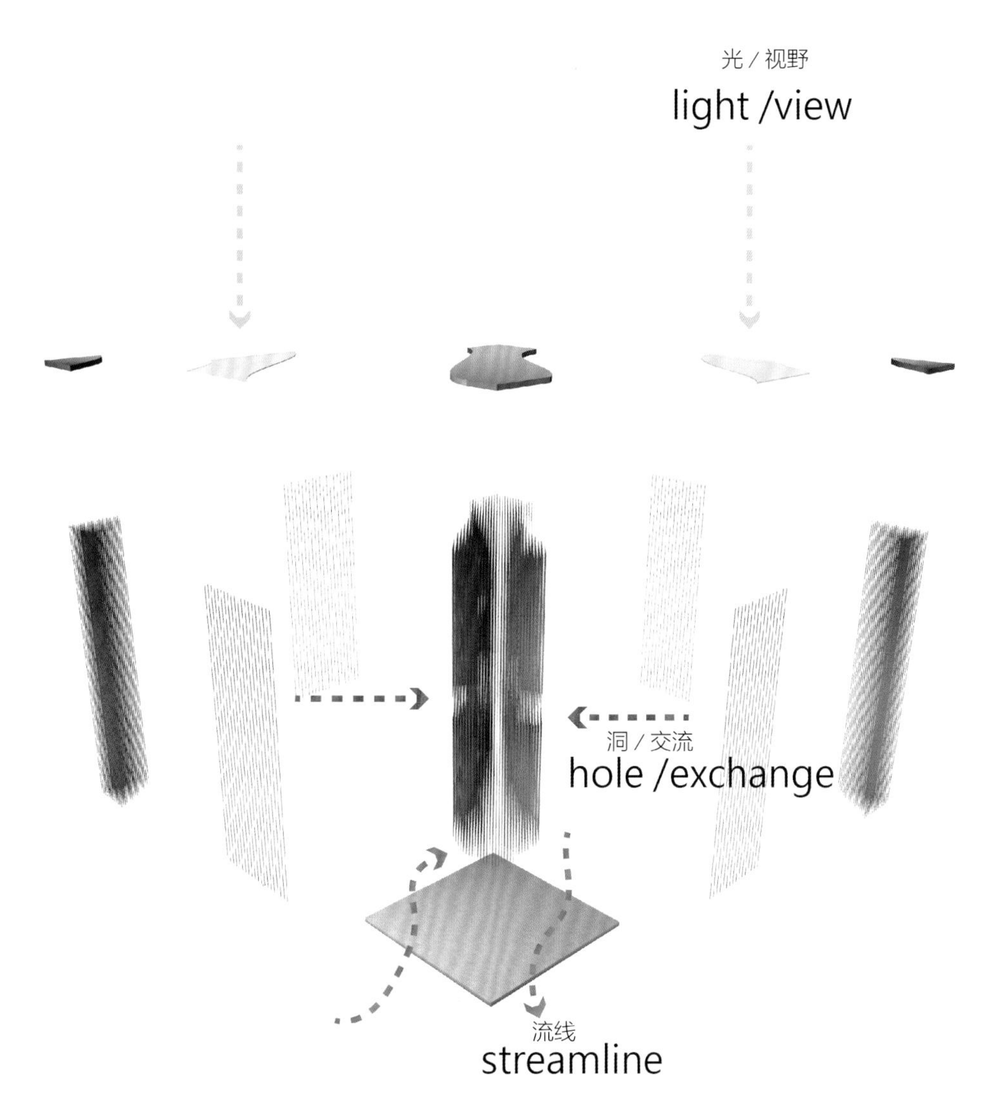

▲ 概念分析图

Concept analysis diagram

突破之外
OFF THE GRID

邓元斌 Jesden Tang（马来西亚）

“看与被看的情况下，只有被看的人才能被体会。”

旁观者清，当局者迷。我们都是社会系统中的一部分，而身处其中，我们真的能看清自己的处境吗?

因此，设计师将此作品命名为“突破之外”，希望观赏者能通过这部作品看到属于我们这个时代的种种因素与现象，也看见我们这个时代正在努力。

“In the relationship between the one seeing and the one being seen, only the one being seen can be acknowledged.”

As the saying goes, “No one is better in his own affairs.” We are all a part of a system. However, can we objectively see our surroundings as they truly are?

Hence, our design installation is called “Off the Grid,” which represents our aspiration that spectators can see the world of our generation better through this design work , and see this generation’s dedication.

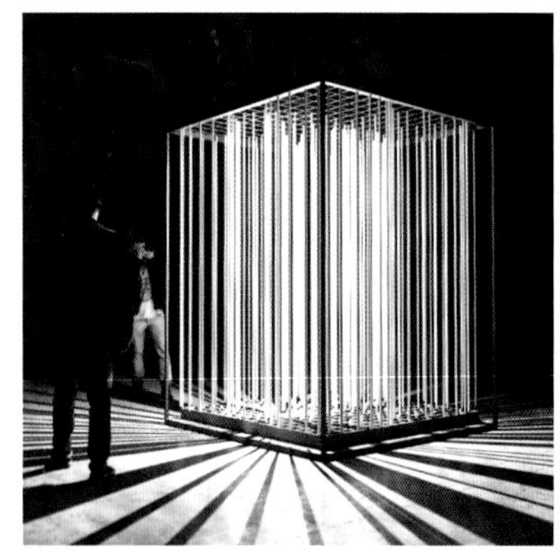

◀《突破之外》效果图
Design sketch of OFF THE GRID

▲《突破之外》效果图
Design sketch of OFF THE GRID

▲ 概念分析图
Concept analysis diagram

格状系统，现代的框架?

格状的造型象征着时代所创造的系统，为人类带来了方便和文明，但同时也带来了局限的框架，限制了人们的思维及行动，在框架下生活，盲目跟随着社会的系统及传统，失去了该有的主见与思维。

Grid System — A modern society that lives in the box?

We used a form shaped by the grid to symbolize a system over time built by the system, which not only brought us civilization and better life, but also limited our thoughts and actions. We blindly follow the system and tradition under the modern society, we have no independent opinions and mindsets at the same time.

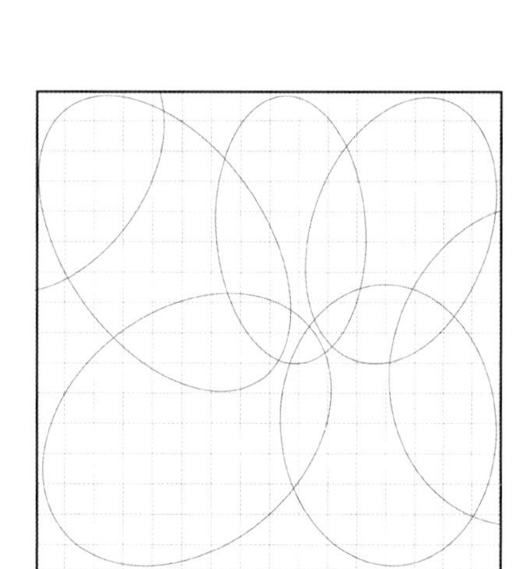

▲ 概念分析图
Concept analysis diagram

领域交叠，边界模糊化。

以设计的角度来看待这个时代，很多关于领域与人的限制或界限正被模糊化，使得现代很多事物都息息相关并互相连接。图形隐喻现在各领域互相交叠，并在框架下共存，形成框架中的秩序。

Overlapping domain and blurred boundary.

Observing the modern artifacts from the perceptions of designers, we can clearly see that the limitations and boundaries between domain and people are blurring, which cause that many things are bonded and connected, a metaphor implyies that domains are overlapping, thriving and adjusting within the grid system.

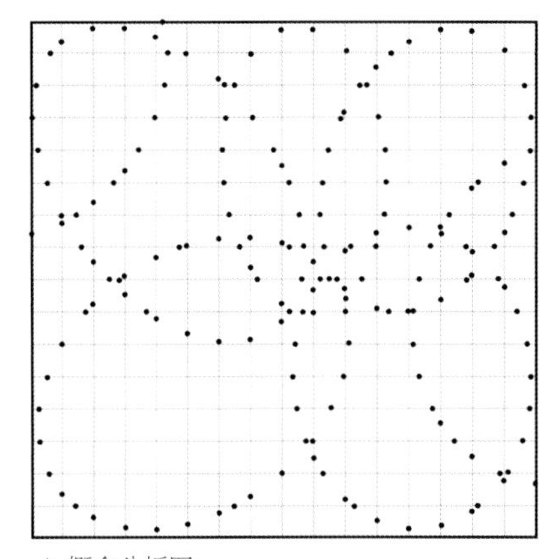

▲ 概念分析图
Concept analysis diagram

领域的虚实，不被发现的真相。

作品中，我们用绳子顺着框架中的秩序点阵缠绕，从立面的角度去看，只看见前后的线条，却不知看似杂乱无章的绳子，其实也在顺着它们所跟随的秩序与规则默默地运行。这也暗喻了现代人看表象而不做深度思考的现象。

The actual domains, the hidden truths.

In the artwork, we tied the ropes to the grid in certain order. From the side views, the seemingly disarranged ropes have specific patterns. This implies that people nowadays only believe things rather being seen than being pondered.

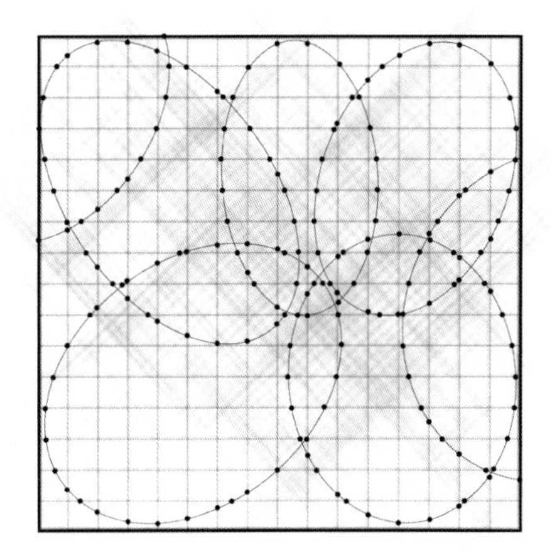

▲ 概念分析图
Concept analysis diagram

突破框架的途径?

如何逃脱这框架? 以物理的原理来说，绳子无法自行脱离框架，就算强制拖出，也会因引力回归原位，绳子就像人，是一种实体。光是一种虚的物质，思想也是，将光比喻为思想，投射在绳子上，才会衍生出影子，这时候绳子才能以虚的形式突破框架，光与影共享共存，无光何来影，无影何来光?隐喻了人必须靠思维才能突破框架，而人与思维的这种关系，是矛盾的，也是共存、共享的。

Escaping from the box?

How to jump out of the box? The ropes in the framed box cannot automatically escape from the box. Even pulled out, the ropes will fall back into the box by gravity. The ropes symbolized people and an entity. People cannot escape the box without assistance. But how about lights? Lights are formless, and so are mindsets. The mindsets were represented by lights, which illuminated on ropes and created shadows. Only this moment can the ropes imaginatively escape from the box. Lights and shadows were of no differences, and this phenomenon implied that people jump out the box by thinking. The relationship of people and mindsets are contradictory, co-existing and sharing.

显与隐

EXPOSE AND HIDE

叶铭盛 Sean Yap （马来西亚）

显身于隐

在这个世界上每个人都有平等的机会去使用各种各样的东西，例如共享汽车、共享空间、共享单车等。显而易见，共享这种行为已经遍布于我们的生活之中了。当然，想要获得与大家共享的可能，你必须先与外界分享自身的某些东西。从这个角度来看，在很多特别的地方使用者有更多不同的体验。举个例子，如果你突然与一个陌生人共享汽车，你一定会觉得“个人领域”受到了这个陌生人的“侵犯”。这时你便会尽力隐藏自己，仅仅暴露自己想要表露的“表面印象”给对方。空间共享也有着相同的情况，因为事实上我们人类或多或少都有私人领地的概念。是的，我们确实共享东西，但是我们共享的不是我们的真实，而是被我们有意识地塑造与隐藏的自我。

大部分的共享空间为了增加人与人之间的接触和互动，都尝试抹去“个人空间”的存在，但是我们从没有认真思考过，这样的共享空间真的能让使用者感受到舒适吗？又或者我们只是突然将注意力转到了处理这个共享空间上？

所以对我而言，大部分共享都是从隐藏或者私密的角度出发的。当你在你自身的区域感受到舒适，想要去分享一些事情的时候，你才会真正地表露你自己。

Hiding yourself before exposing yourself

In this world, everyone stands an equal chance to use everything, such as car sharing, space sharing, bike sharing. It is clear that the idea of sharing is omnipresent in life. Before being shared, you have to present something. Mostly, users and owners have different experiences to the same thing. For an example, if you share a ride with a stranger, the emotion of “self-defense from being violated” make you eager to conceal yourself. Then you try to hide yourself, and only expose the “available impression” to others. Space sharing is analogous because most people accept the idea of personal space. We do share, but we normally share a part of ourselves.

Most of the sharing spaces focus on exchange and ignore the existence of “personal space.” Do people really like space sharing, or it represents a distraction?

For me, most of the sharing is based on the hidden or personal perception. When you are comfortable in your zone and want to share something, you would expose yourself.

▲《显与隐》效果图
Design sketch of EXPOSE AND HIDE

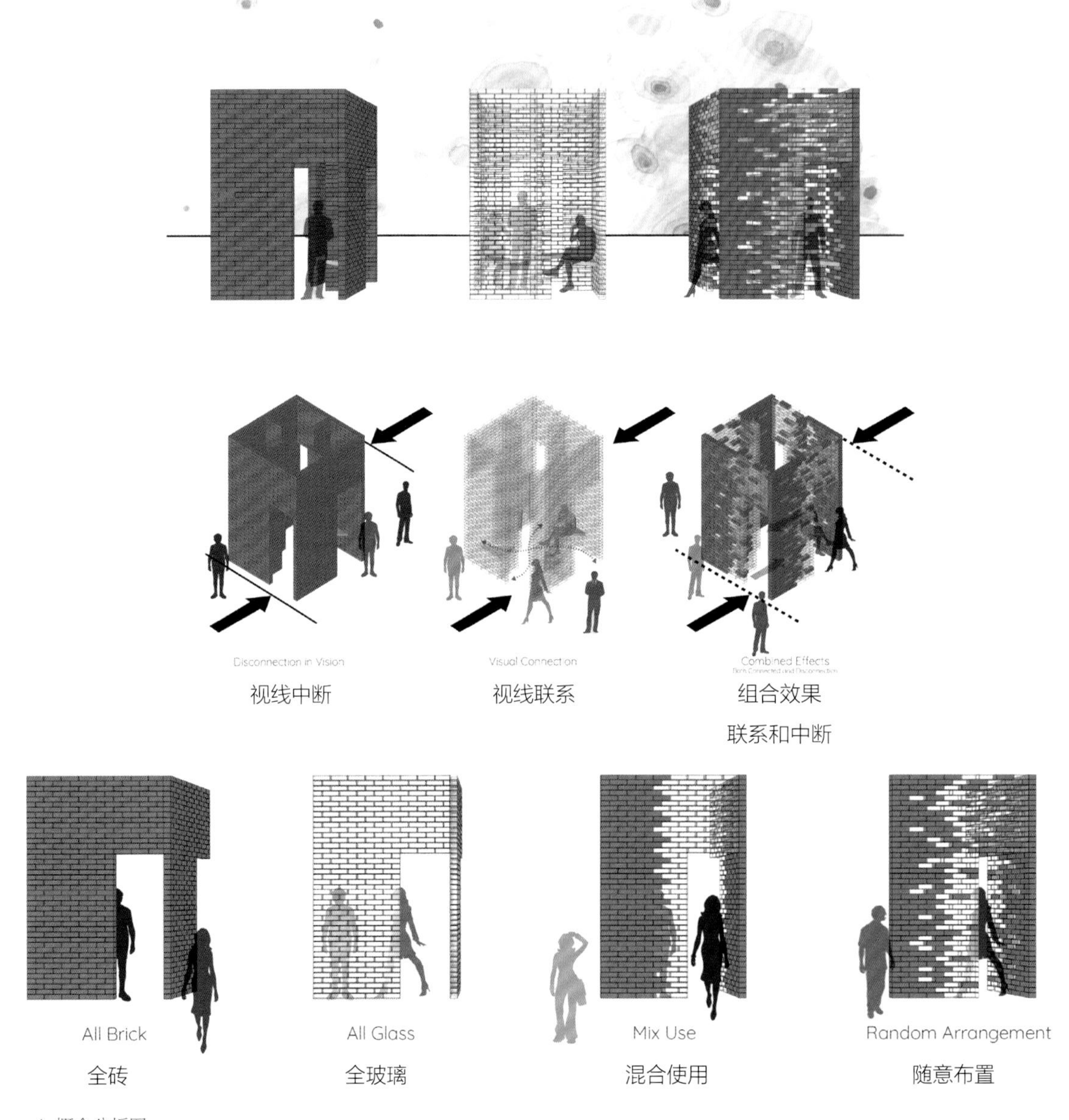

▲ 概念分析图
Concept analysis diagram

砖排列：

试图模糊显和隐的差异，但仍然保有这两者的特征。

Brick Arrangement

Trying to blur the differences between exposing and hiding without compromising the distinct features.

Roof System 屋面系统
Glass Brick 玻璃砖
Red Brick 红砖
Concrete 混凝土
Wall System 墙壁系统
External Wall 外墙
Divider Wall 隔墙
Floor System 地板系统
Benches 板凳

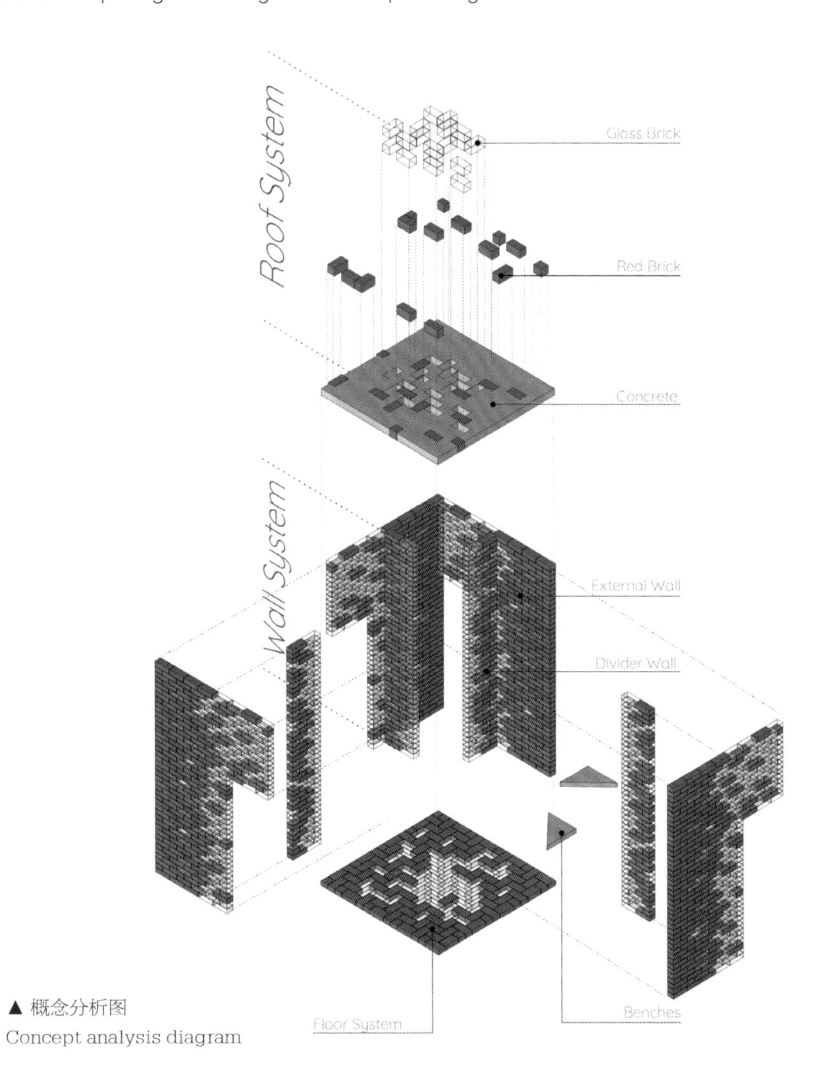

▲ 概念分析图
Concept analysis diagram

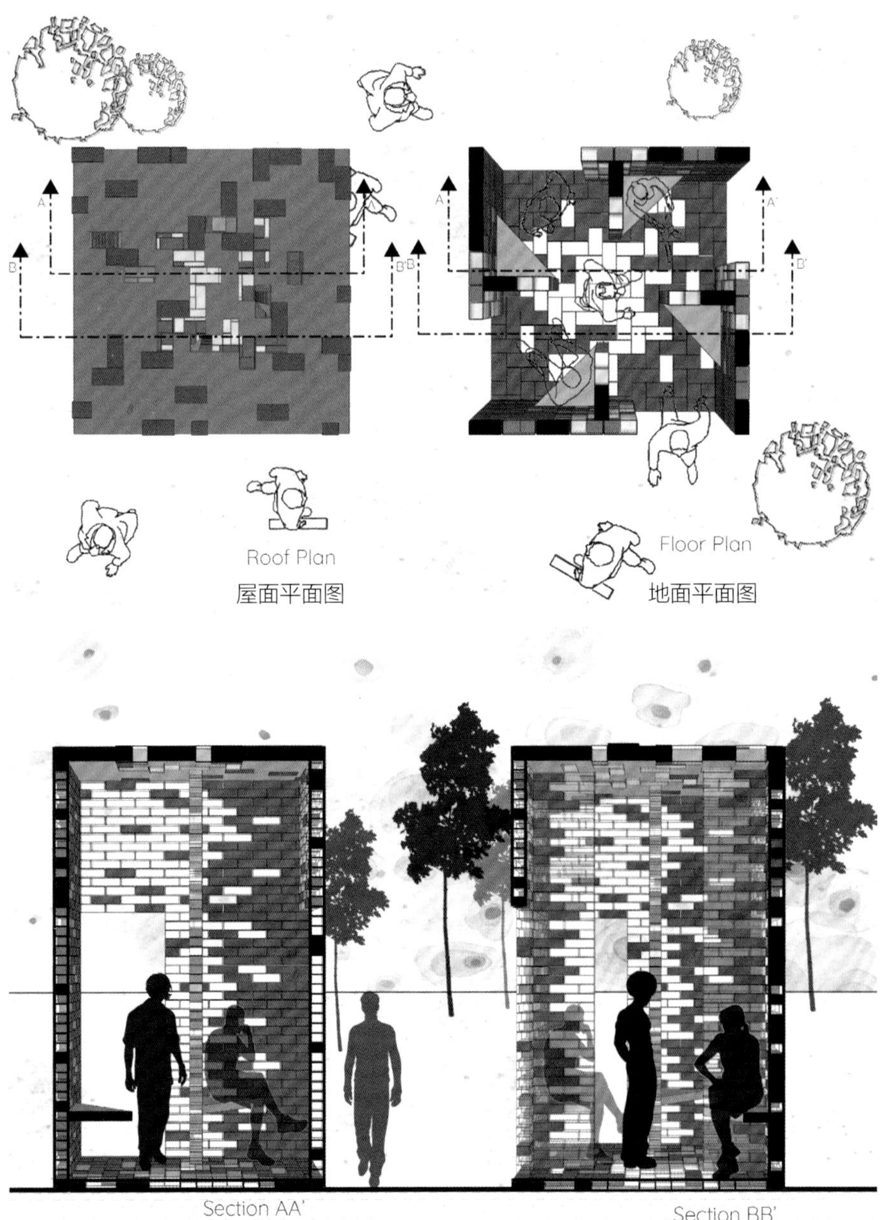

◀ 概念分析图

Concept analysis diagram

▲ 效果图
Design sketch

平面和细节：

砖系统的语言延伸到屋顶和地板系统。红砖包围着玻璃砖，并且运用相同的系统模糊了两者的边界，在中间创造了一个轻松、开放、集中的区域。

这使我们能够在一个共享空间内保留一些私人空间。鼓励人们运用角落区域，以他们的意愿分享他们的想法。

Plans & Details

The brick arrangement system was applied to the roof and floor systems. The glass brick was surround by the red brick, and was using the same system that obscured the boundaries, and created an amiable, open and centralized area in the middle.

This idea allowed us to have both sharing and personal corner spaces, and encouraged people to utilize personal spaces and share opinions.

图书在版编目（CIP）数据

共·享 ：设计师的人文思考 / 邵唯晏，卜天静编著 . -- 南京 ：江苏凤凰科学技术出版社，2019.1
ISBN 978-7-5713-0118-7

Ⅰ. ①共… Ⅱ. ①邵… ②卜… Ⅲ. ①室内装饰设计-研究 Ⅳ. ①TU238.2

中国版本图书馆CIP数据核字(2019)第025330号

共·享——设计师的人文思考

编　　著　邵唯晏　卜天静
项目策划　凤凰空间／杜玉华
责任编辑　刘屹立　赵　研
特约编辑　杜玉华

出版发行　江苏凤凰科学技术出版社
出版社地址　南京市湖南路1号A楼，邮编：210009
出版社网址　http：//www.pspress.cn
总 经 销　天津凤凰空间文化传媒有限公司
总经销网址　http：//www.ifengspace.cn
印　　刷　北京博海升彩色印刷有限公司

开　　本　710 mm×1000 mm　1／12
印　　张　13
版　　次　2019年3月第1版
印　　次　2019年3月第1次印刷

标准书号　ISBN 978-7-5713-0118-7
定　　价　59.80元